THE ESSENTIAL BOOK OF
TIME

THE ESSENTIAL BOOK OF
TIME

Master the Mysteries of Time
in **12** Short Chapters

STEN ODENWALD

ARCTURUS

Dr Sten Odenwald received his PhD in astrophysics from Harvard University in 1982 and has authored more than 100 papers and articles and more than 20 books on astrophysics and astronomy education. He is a frequent contributor to *Astronomy* magazine on topics related to cosmology and heliophysics and is an active member of the NASA Heliophysics Education Consortium. He has a number of websites promoting science education, including his blogs and other resources at 'The Astronomy Café'. He has also appeared in National Geographic TV specials, as well as several BBC programmes such as the Curiosity Stream series on space weather, which debuted in 2019.

Other books by Sten Odenwald include:
A Degree in a Book: Cosmology
Knowledge in a Nutshell: Astrophysics

Discovering the Universe
The Story of Quantum Physics
The Essential Book of Black Holes

ARCTURUS

This edition published in 2025 by Arcturus Publishing Limited
26/27 Bickels Yard, 151–153 Bermondsey Street,
London SE1 3HA

ISBN: 978-1-3988-4771-2
AD011372UK

Printed in China

CONTENTS

'...IT'S YESTERDAY ONCE MORE'

Why does time exist? What are its innermost features? Why is it that no two clocks in the entire universe, no matter how close, measure the same time passing? These are the questions being probed by 21st-century physicists.

We all seem to know what time is. That's because this knowledge is almost literally written in our bones. We don't have to give it a second thought. Since a very young age we live our lives 'by the clock' to set up meetings, plan our daily activities, do our fitness workouts and many more things. Against this rock-solid framework of time that seems to pass of its own accord, we have the internal time we experience through our mental activities both conscious and unconscious. This internal time-keeping is malleable, with some moments passing slowly while others zip by and are gone. Where does our internal time-sense come from? Why do we feel that we have existed for a long time, and that future events seem to flow towards us like a river? The answer to these questions takes us into a fragmented world where events in subjective time are not as well ordered by cause and effect as we might imagine!

One of the truest comments about the human experience of time in recorded history was uttered by Saint Augustine (354–430) who said, *'What then is time? If no one asks me, I know what it is. If I wish to explain it to him who asks, I do not know.'* This is the essential paradox that time confronts us with. We can populate space with fixed markers, be they stones or houses, that tell us where we are. But time remains unmarkable. There are no reference points to mark time that are independent of some event such as a birthday or a sunrise. Unlike space, where we can look at distant and near objects out to the horizon as they mark intervals in space, if we are alone in a room and remain motionless, the room dissolves into a timeless 'thing' until we take the next breath or can sense our heart beats. Yet over the millennia, humans have developed a manic reliance on the measure of time without really understanding what it is. First there was the passing of the seasons that guided ancient agrarian societies to plant and harvest. Then the regular phases of the moon led to a 30-day interval that subdivided the seasonal year into 12 or 13 'months'. Finally, the individual day was subdivided into hours and even finer measures as religious rituals demanded specifically timed observances. But this only covers our conscious uses of time to organize our activities. Time and its measure have impacts that are far broader than the interests of one species.

Time is a phenomenon of nature that has been known to us in our very bones since before our species appeared on

Humans have been keeping track of time since before recorded history, tracking the sun through its seasonal movements such as the spring equinox at Stonehenge.

this world. Every organism, every cell, has a sense of time encoded into its very DNA. Bacteria that use photosynthesis are aware of the 24-hour cycle of day and night, but non-photosynthetic bacteria such as *Bacillus subtilis* also have this sense as well as temperatures rise towards noon and fall towards nighttime. Bacteria ingeniously synchronize the ebb and flow of their molecular workings to the diurnal cycles in sunlight or in temperature. Some have even evolved photo receptor spots, primitive eyes to make this synchrony more precise.

Animals and plants also understand the passing of the seasons. Animals time their migrations to the cycle of abundance in the spring and summer in one locality, and move to other localities where plants are available during autumn and winter. Birds often migrate from hemisphere to hemisphere to live in a perpetual spring and summer as the deathly cold winters pass below them. Lacking a brain and neurons to think with, plants nevertheless 'know' when to shed their leaves in the autumn, and grow new ones in the spring.

But measuring time accurately is not the same as understanding why it exists in the first place. To go beyond simply recording time intervals with even more precision, we have to devise a set of ideas about its nature and properties that we can observe and put to the test. The best way we know for devising explanations for aspects of the physical world is through the Scientific Method. The basic idea is that we propose an explanation for how something works (theory of

special relativity), which predicts that certain things should occur regularly in the form of a law ($E = mc^2$). We test this new law by making observations. A theory stands or falls on its ability, not only to explain known observations and laws, but to predict new ones that in turn stress-test the

validity of the theory. For example, Isaac Newton's theory of universal gravitation explained why Johannes Kepler's three laws of planetary motion existed, but also enabled astronomers in the 18th and 19th centuries to predict and discover the new planets Uranus in 1781 and Neptune in 1846. Flaws were eventually found in Newton's theory of universal gravitation and it was replaced by Albert Einstein's theory of general relativity in 1915. This new theory of gravity went on to explain deviations in the orbit of Mercury, the existence of black holes and gravity waves; none of which were predicted by Newton's theory. Currently, physicists are attempting to find errors in Einstein's general relativity that will let them create an even better theory of gravity in the future. This approach, using the Scientific Method to explore the nature of time, is taking us on an amazing ride from brain research to quantum theory and beyond.

Does time flow like a river? Some say you can never place your foot in the same place in the (time) stream twice. This idea was proposed by Heraclitus of Ephesus (530–470 BC).

YOUR BRAIN ON TIME

Neurons connected together through their dendrites and synapses create complex local circuits including those responsible for our perception of time.

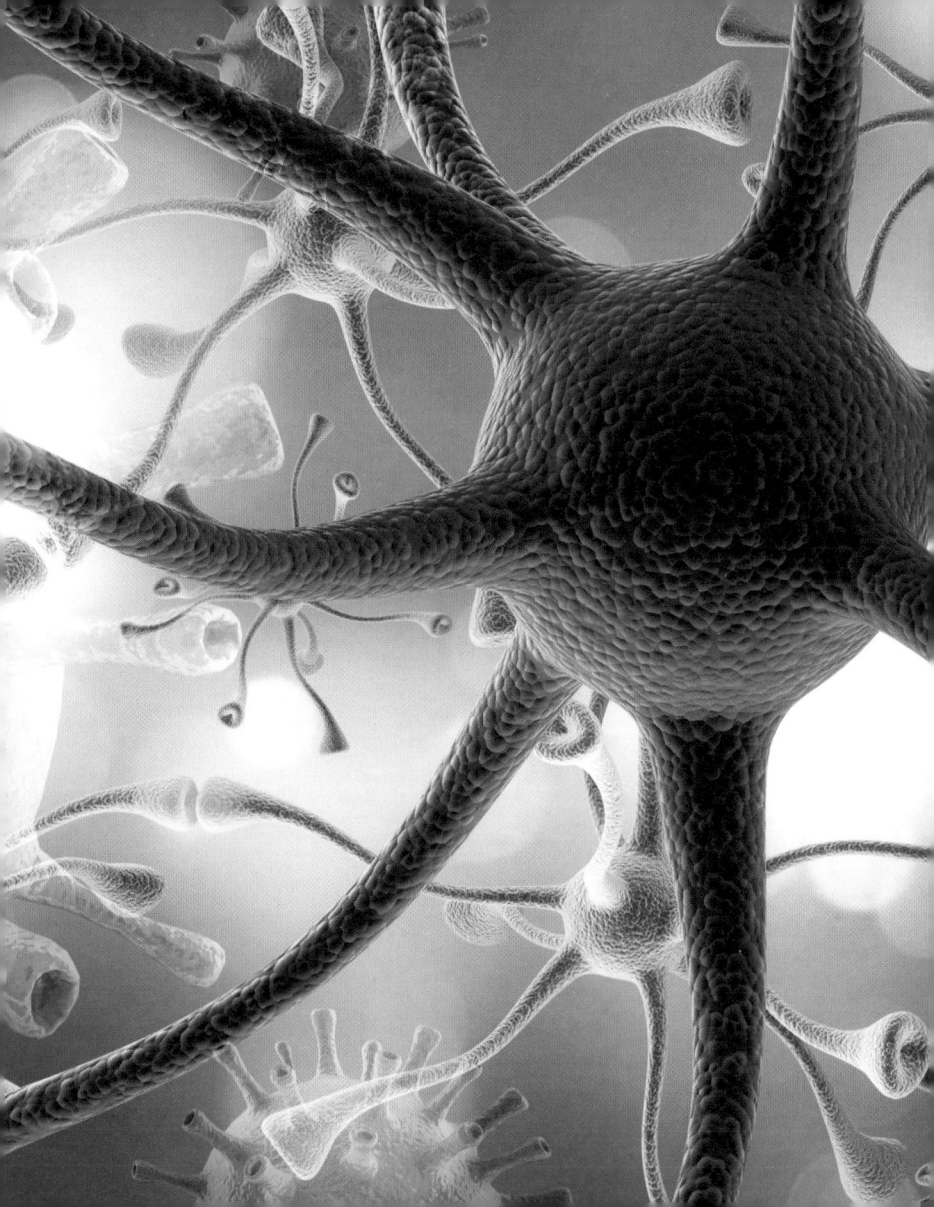

How does the brain do it? How does it take in gigabits of information from its many senses and knit them together into a single unified experience? The answer is that our brains contain colonies of neurons and their dendrites, which connect to each other to form circuits. These circuits also connect with other colonies of neurons far away to provide inputs to their operation. It is a bewildering network that resembles how computers are connected around the world. There are local area network 'nodes' within homes and office buildings. These nodes are connected via the World Wide Web to other networks to share e-mail and to transfer files. Each of these nodes performs its own independent actions and gathers its own local store of data from which to base decisions. One might be running a program to model some aspect of climate change. Another node might be an

A specific sensory stimulus enters our consciousness when the locus of neural circuits processing the stimulus trigger an avalanche of activity among far-flung circuits so that the entire brain becomes temporarily excited.

element in a global multiplayer game. A third node might be writing an article that will be uploaded and shared across the internet to inform other interested parties. Like brain circuitry, the nodes operate independently; taking in data and providing an output shared through the network. Sometimes, however, some major event such as an earthquake, a war, or a major scientific discovery stimulates thousands of these distributed nodes. The number of nodes passes some threshold and portions of the World Wide Web flash into operation at the same time as millions of people share their

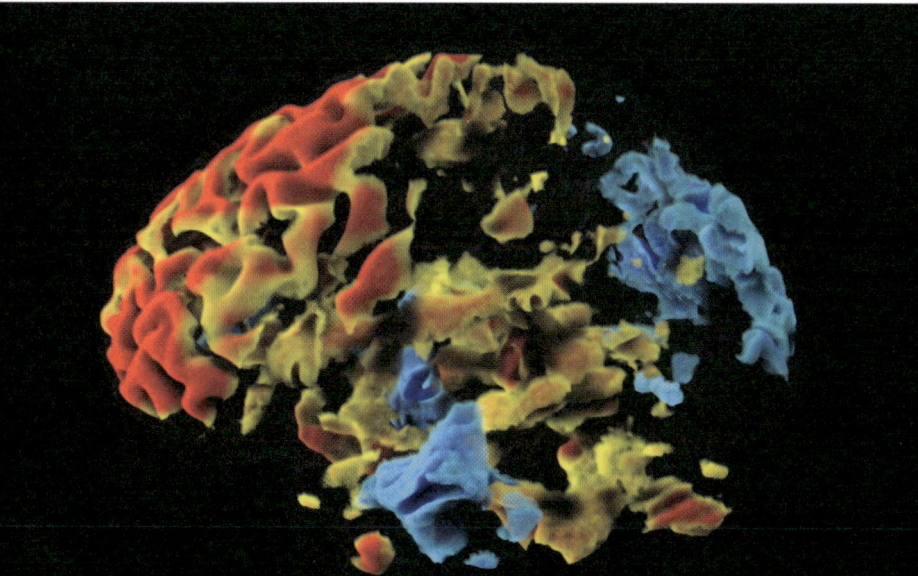

experiences of the event. For our brain, this event represents the moment when a collection of inputs to specific local brain circuits operating unconsciously, passes a threshold where much of the cerebral cortex is now involved in a conscious experience.

The first thing the brain has to do is to have a sense of its own body and how it is located in space. It also has to identify this 'self' as being different from that of other people. If it cannot do this accurately, it cannot decide how to move in space, anticipate the consequences of that movement, or how to anticipate and empathize with the actions of other people. Nearly all of this activity seems to be relegated to a single area in the brain called the temporoparietal junction. This brain region takes information from the limbic system (emotional state) and the thalamus (memory) and combines it with information from the visual, hearing and internal body sensory systems to create an integrated internal model of where your body is located in space. The temporoparietal junction tries to create a coherent body image from many different, and sometimes contradictory sensory inputs. When this process breaks down because the contradictory information is too strong to inhibit or ignore, you experience that you actually have two distinct bodies in space. This seems to be the direct, neural basis for out-of-body experiences.

But there is an even stranger brain region that decides where the body and self ends in space, and where the outside world begins. Called the posterior cingulate body, it plays a huge role in self-location and body ownership. Another region called the posterior superior parietal lobule gives us a sense of the boundary between our physical body and the rest of the world. When activity in this brain region is reduced, the individual seems to lose a sense of where their body ends and the rest of the world begins. The feeling is one of having 'merged with the universe' and your body is in some way infinite. Mindfulness practices such as meditation can modify the stimulation of this region and give the practitioner exactly this dramatic experience.

Once the brain has partitioned its experiences of the world into 'outside' and 'self', it can now get on with the business of processing the sensory data into a coherent snapshot of what is occurring around the self and how these circumstances affect its survival. The flow of sensory information at any one time is enormous. It starts out as separate streams of information that flow to specific brain regions and neuronal circuits as their first stop, but then after that, the information

radiates to many different regions in the brain where it gets mixed with our emotions, and even with other sensory information. This is a process that is called association. It is such a complex process that a large volume of the brain called the association cortex is dedicated to this function.

The brain creates meaning by associating sensory information with specific categories of past experience. Nothing can really be understood except through a complex process of stimuli being associated with other things you have experienced, or learned. It's like some enormous, interlinked tapestry of connections. Adding a new bit of information, whether it is factual or not, is always about fitting it into what has already been experienced in some way. But if that were all that there was to our experiences, we would simply be walking encyclopedias. How do we create models of the world that let us function and survive without having accidents and getting killed all the time? There are at least two basic ways that we create associations. The first is associations in space. The second is associations in time.

Associations in space include recognizing objects like chairs, trees, cars and people. The reason this works so well is that we live in a world filled with many different kinds of more-or-less fixed objects so that two or more people can agree they have similar attributes. A visual cortex 'image' of a cat becomes associated with the auditory sound of purring, and the tactile sense of fur. Broca's Area is responsible for language production, writing and speaking, so that the word 'cat' is now added to the string of associations we form mentally. Finally, in Wernicke's Area our brain 'understands' what is meant by the written or spoken word 'cat' in the totality of mental impressions we have of it. Meanwhile, the three sensory channels, sight, touch and hearing, are linked to the limbic system where they are tagged, not with a fight-or-flight reaction but an impulsive reaction to 'like' the cat as a non-threat.

Patterns in space let us recognize the many different kinds of objects that fill our world. In the association cortex, once these identifications have been made, they are accessible to the language centres where they are tagged with words that can be spoken or read. Once this step happens, two individuals can have a meaningful conversation about the world beyond their bodies that the senses can detect. Information is collaboratively shared. For instance, you might have seen a banded snake (scarlet kingsnake) in which the black and red bands are adjacent and you know this snake is harmless and can be held. Then you meet another person

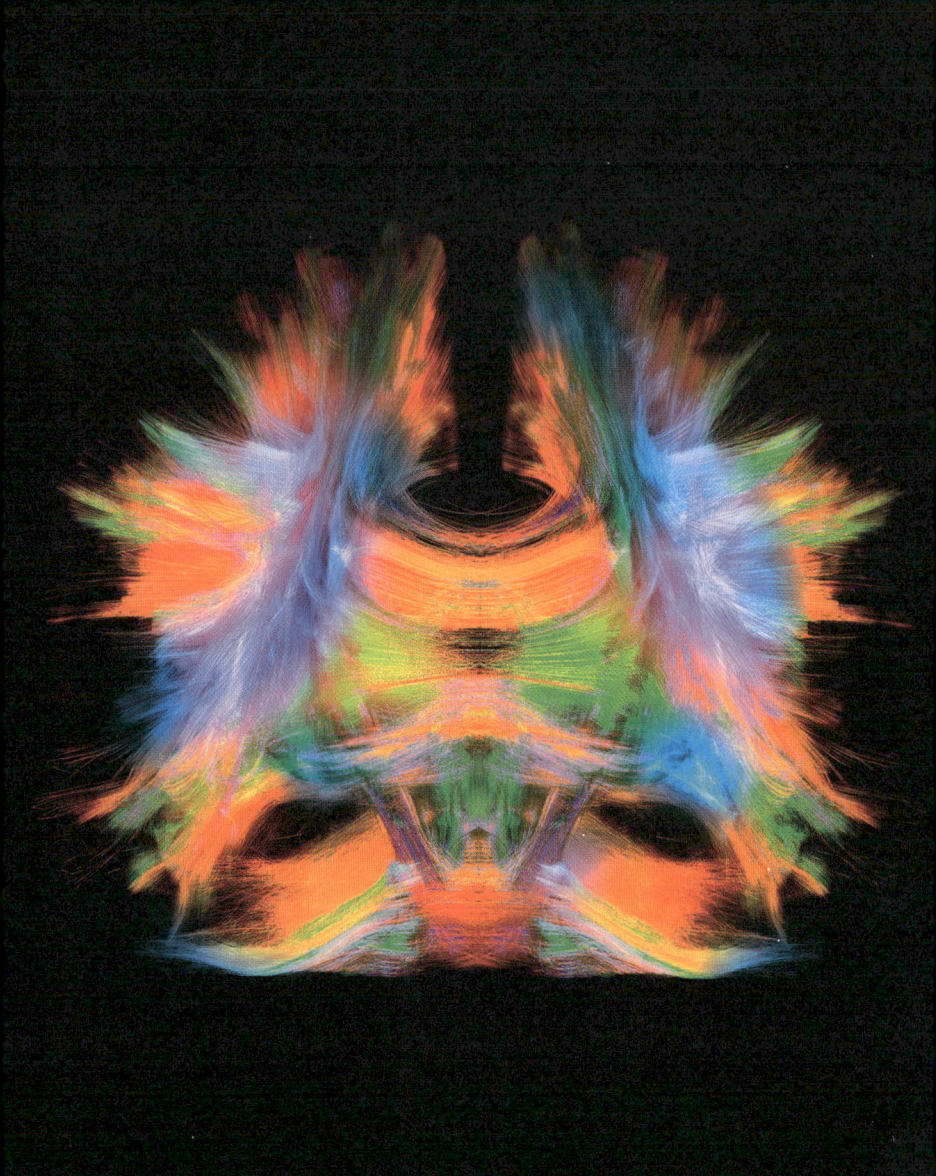

that sees a banded snake (coral snake) in which red and yellow are adjacent and he tells you this one is poisonous and should be avoided. Together you come up with a new rule '*Red-touch-yellow: kill a fellow. Red-touch-black: friend of Jack*' and teach this to your children. This sharing of information outside the self is a major advantage for self-preservation. You learn about things that help your survival without actually having had the experience yourself. The next step in the association process is even more interesting and here is where time comes into the picture.

Some of the brain connection pathways identified by the Sherbrooke Connectivity Imaging Lab, some of which help to interpret sensory data as patterns in space and time.

THE AMBER OF THE MOMENT

A moment frozen in time. A fly trapped in amber.

A century of accumulated knowledge from stroke victims, brain pathologies and direct neural imaging has revealed the many specific brain systems that have to work together to provide us with a sense of time. These include the prefrontal cortex, basal ganglia, and anterior insular cortex. The most surprising feature of our time-sense is that it is not based on some internal master clock that counts a sequence of moments with clock-like regularity. This is remarkably parallel to the fact that human vision is not created by some 'homunculus' inside our brain looking at the equivalent of a television screen. Each cell in our body has its own chemical clock controlled by its own DNA and the transcription and expression of this DNA to form proteins. As the concentration of a specific protein increases, it reaches a critical concentration where it inhibits the further expression of the protein. The concentration then decreases until it reaches a second concentration threshold where it once again triggers the transcription of more protein and the cycle repeats over the course of about 24 hours. These cellular clocks determine the pace and organization of cell division, but collectively do not add up to a global, unified clock related to our sense of time. In fact, there does not seem to be any correlation between the circadian 24-hour rhythm and the pace of our temporal judgements over time spans from seconds to a few minutes. This seems to suggest that we have brain circuits that operate in the time domains of a few seconds, a few minutes and hours but are distinct from each other in how they operate and react to the outside world.

Just as the brain generalizes a collection of special associations to define the concept of 'cat', it can detect patterns in time in the world beyond our physical sensoria (e.g. eyes, ears, etc.) and begin to see how one event leads to another as a rule of thumb or a law of nature. We do not describe patterns in space the same way we do patterns in time. A pattern in space is simply a detailed static description of what something looks like or its immediate properties such as shape, colour or mass. But a pattern in time can only be described by what we call a story or a history. This is a sequence of spatial descriptions ordered in time that tells us something very basic about our world.

Instead of being a random collection of events like static pictures in a photo album or furniture occupying our living room, our physical world contains a basic collection of rules

that follow a 'logical' *If A happens then B happens* pattern in time, which is the basis for what we call causality. If I drop a stone off a tall cliff, it will fall downwards to the valley below. If the sun rises and sets today, it will do so again tomorrow. There are many such patterns of events in time that reoccur with such regularity that they form their own category-in-time much as 'cat' and 'chair' did in the space context. *'If I visit a waterhole with lots of animals, there is a good chance that lions may also be present.'* This ability to discern cause and effect seems to be localized to activity in the cerebellum and the right hippocampus.

Over the centuries and millennia, the patterns in time we have been able to discern about the outside world have become so numerous that we have to put our children through longer and longer training periods to master them. We can point to a tree and say 'tree' but we can't point to an object and say 'law'. A law has to be presented in the form of a story that unfolds its character in time, and it takes a lot longer to tell a story than simply point at an object and say 'tree'.

It is estimated that, among all the neurons and dendritic interconnections, our brains could store about 2.5 million gigabytes of information. But the brain only presents about 50 bits per second to the conscious mind. All the rest of this information is being processed by your unconscious mind. The amazing thing is that with this degree of compression, our senses still give us the impression of being embedded in a high-definition visual world and a fully detailed acoustic sensorium. This massive compression of information also defines our perception of time, which is highly focused on the conscious experience of Now lasting about three seconds or less.

There are two Yous involved in experiencing the world. The unconscious You operates simultaneously with the conscious You but does not have to waste any time telling the conscious You what it is doing. It is busy keeping your heart pumping, processing your retinal and auditory data, making snap judgements about how you are feeling both physically and emotionally. If you are a gymnast in flight, the unconscious You guides your body through complex muscular movements in 3D space. You have no conscious sense of this massive data flow upon which your brain is making judgements and most importantly anticipating what is in store for you in the next few seconds. Occasionally, however, it passes information up into your conscious You once some stimulation threshold is reached because the particular collection of stimuli it is receiving are novel and

unexpected. You can sit in a noisy room filled with people talking and not be paying attention to the murmur of voices, but as soon as your unconscious You detects your name being mentioned, your conscious attention immediately activates and you focus on the direction towards which you heard your name.

While your unconscious You is processing information at cadences of milliseconds, the conscious You is only aware of visual information that changes every 20 to 30 milliseconds. Your ear is better than your eye in this respect because you can consciously follow the changes in pitch and loudness in a song that take more than about 0.05 milliseconds (lower than 20,000 Hz). The slowest and fastest musical tempos range from 40 to 200 beats per minute but the brain can replay events at higher speeds. The brain is also more than capable of telling time in the sub-second range. Our brains can even synchronize what we are hearing with bodily motor movement in the form of dancing. Most animals, by the way, cannot reproduce this simple human ability to 'keep a beat'. Visually, when we are presented with two consecutive still frames from a movie, we can discern that they are presented separately if the time interval between them is about 30 milliseconds or longer, but with a shorter cadence they blur into a single continuous event that includes the illusion of motion. To create the illusion of a movie, the brain has to synchronize multiple sensory streams into a single coherent story and internal model. The window of time where these sensory events appear to be from a single moment is about 100 milliseconds. What this means is that any sensory input to consciousness that arrives within this span of time will be perceived as simultaneous. Amazingly, within this 100 millisecond window, sensory information can arrive randomly and you will not notice it. For example, you watch someone bouncing a basketball on the pavement as they walk away from you. Light travels almost instantaneously but sound travels at 340 m/s. By the time the person is about 34 metres away, you will start to see the sound of the ball hitting the ground lag behind the image of the contact moment. The brain has to do a lot of computation and passing sensory information along different association pathways and circuits. This 100 milliseconds seems to be how long it takes for the brain to 'cook' the data and give you the feeling of the essence of your Now moment. But the brain doesn't stop there.

How do we go from Now to the all-encompassing feeling of having existed far beyond this brief moment? Your brain

generates your sense of the past only through its stored memories. The cerebellum, prefrontal cortex, basal ganglia and the anterior insular cortex are all involved in integrating the sensory information into a model of the world. This also has to happen within this 100 millisecond window. Once it has created an internal model or 'image' of what is going on Now, it has to immediately relate this Now to the previous Nows. Some of these previous Now impressions are still temporarily in short-term memory, but they quickly pass into long-term memory, especially if your limbic system has tagged that Now as being noteworthy for your survival or other issues. This is handled by the hippocampus, the frontal lobes, the median temporal lobe and the parietal lobes. Because of the finite speed of nerve impulses and the many synaptic connections, the information we think we have at our immediate conscious disposal in the Now is already as much as 300 milliseconds out of date. If this is what it is like to see your Now in the context of your past, why do you also feel as though there is an equally real future up ahead to which Now is part of a continuum of events?

Your brain's primary task is to find patterns in the world that you can act upon to preserve your life. You draw on memories of past, similar experiences to gauge how you should behave in this particular Now. At the same time, your brain is doing another remarkable thing. It is planning for the future to maximize its survival. Part of this planning is unconscious and part is conscious. Your sense of the future comes only from predictions that your brain generates on-the-fly about what will happen in the next few milliseconds, minutes or hours.

A pianist trains their cerebellum to automatically flex particular fingers at specific moments to play a complex composition consisting of 20,000 notes. The art of music is to train your brain consciously how to do this finger-flexing, but the ultimate goal is to make even the most complex sequence of flexings entirely automatic and unconscious. You do not want the slow brain to be involved, instead you want to operate like an automaton with only the fast millisecond timings of individual neurons to guide the music. This is why practising various scales on your instrument is so vital to the process, although out of necessity it has to be repetitive and boring. You are forcing your brain to optimize brain circuits to operate without your conscious control to simply play a sequence of notes seen visually on a piece of paper without consciously thinking about where your arm, hand and even fingers should be over the

keyboard from instant to instant. The British composer Kaikhosru Shapurji Sorabji wrote a phenomenally difficult musical score called the *Opus Clavicembalisticum*, which he describes thus: '*The closing four pages are so cataclysmic and catastrophic as anything I've ever done – the harmony bites like nitric acid – the counterpoint grinds like the mills of God....*' It was written in 1930 for solo piano and takes over four hours to perform.

Similarly, your brain constructs one internal model after another as each Now passes. It unconsciously knits them into a tableau of what just happened and what is about to happen and passes this to the conscious You. You then experience Now as fully integrated into a momentary timeline of your experience. In some limited sense, past and future are just mental constructs that your brain uses to organize its sensory experiences. Amazingly, our brains re-use some of the circuits for spatial pattern organization to build this sense we have of time. The result of this sharing is that we spatialize time into a timeline that we can draw on a piece of paper, or some internal mental image of a spiral or a circle that we navigate day by day and year by year. We also incorporate spatial references to time intervals. For example, '*I haven't seen you in a long time!*' or '*I live ten minutes from here.*'

A complex score for music that requires a performer to execute hundreds of key strokes per minute on a piano.

What happens when you are no longer able to relate one Now to the next? You are stuck in a world of isolated impressions built upon a constantly changing present

moment in time. This extreme affliction is called Korsakoff's syndrome. The neuropsychologist Oliver Sacks described many cases of this malady in his book *The Man Who Mistook His Wife for a Hat*. A particular patient who could not remember events for more than a few seconds constantly babbled as he tried to verbally create a story for himself that knitted together what he was experiencing. It was an incessant process from the time he woke to the time he went to sleep. As Sacks puts it: '*He continuously has to erect bridges to create a story to fill in the growing gaps in who he is at any moment.*' This condition shows that there are in fact two kinds of time experiences. Investigators call these prospective and retrospective timing.

Prospective timing is a true temporal task in that it relies on the brain's timing circuits. It fixes the Now of your experiences to a short span of events from which you can create a brief story such as 'I just woke up' or 'That donut tastes sweet.' It does not rely on connecting Now to any long story you might have about yourself or deep memories in time that tell you how you may have responded to the events of the current Now. It is the world that the Korsakoff's syndrome patient inhabits. Retrospective timing, on the other hand, is the brain's attempt to infer the passage of time by reconstructing events stored in memory. In the absence of this ability to form new memories you will be trapped in an infinite loop of an unchanging present Now. You won't be able to make sense of where you are or how you got there. The only sense your brain can make of your present moment is that you just woke up because you have little or no memory of what happened in the previous minutes and hours.

How does the brain store these 'stories' that literally are the threads in the tapestry of our conscious lives? A brain malady called hypermnesis was investigated by Oliver Sacks and also described in his book *The Man Who Mistook His Wife for a Hat*, with some remarkable discoveries. If you artificially stimulate a point in the cortex of a patient with this condition, the patient will immediately have the experience of a string of events that form a part of one of these stories. The patient begins to describe the story in great detail until it reaches its end. It's like randomly putting a phonograph needle down upon a record and letting the record play for a few seconds. This led Sacks to propose that story memories are stored among a specific cascade of neural and synaptic connections like programs stored in a computer waiting to be executed. This means that our story memories are stored 'iconically' much like an application

on your smart device. When activated, the memory 'app' unfurls its events as a sequence of experiences in time. So, our memories are stacked like records in a record collection and can be reactivated by intentional recall; by a traumatic or emotional accident; or by a jolt from an electrode.

All of these memories of Nows unfolding as stories you can recall are also built into your internal model of the Outside world and how you should operate to maximize your survival. We have the unconscious You building and updating these models, and the conscious You acting on these models through 'hunches' and 'expectations' of what might happen next. But the conscious and unconscious You share many of the same neural networks, which gives us the visceral experience of time being fluid. Who or what is it that is looking at your brain model to create your experiences of a time-ordered consciousness with free will? The answer seems to be that, just like the experiences of 'red' or 'sweet', the experience of being conscious is also just such a feature that neuropsychologists and philosophers call a qualia. Amazingly, it seems that your 'sense of time' and your 'sense of being conscious' are nothing more than You seeing a series of Nows from inside the movie projector, and not experiencing them like a movie-goer in a theatre watching the screen.

While the brain is the origin of our internal sense of time, it exists within the vaster physical system of the universe. In this universe, there do appear to be basic laws, forces and forms of matter from which all of our physical experiences are fashioned including the substance of our brains and the very operation of the neurons themselves. Scientists have been collaboratively studying this world outside the brain for over 500 years to identify 'universal' features that persist no matter who observes them. These many observations, with increasingly more sophisticated measuring tools, have led to a small number of theories that seem to explain nearly everything that we know how to observe and measure today. The two largest of these theories are quantum mechanics and general relativity. Each of these have something important to say about the nature of objective time outside the cerebral cortex.

TIME GOES BY...SO SLOWLY

An untrained sky diver experiences a terrifying sensation of time passing in slow motion as the prospect of death looms.

For humans, time as we experience it is not a series of intervals of fixed duration. Time seems to have a plastic character to it that passes at a speed that depends on many internal circumstances. It tends to speed up when we experience intense emotions of fear or love, and slows down as we age. We know that our brains create the perception of time through a variety of complex associations of the current Now with past memories and future predictions. So, there must also be some feature of how our brains work that elicits this very curious change in the pace with which time seems to flow. This topic is actually a very active area of brain research. Some research seems to claim that subjective time does in fact slow down, while other tests of this effect can find nothing significant going on. For example, Chess Stetson and his colleagues at the California Institute of Technology set up an experiment to test this effect. Volunteers jumped off a 50-metre platform into a safety net and looked at a wristwatch with an LED display. The display was programmed to show a sequence of numbers in rapid order that could not be discerned as unblurred under normal time. They all experienced the time dilation effect. However, their time discrimination measured by the wristwatch displays showed no dilation in time in the lengthening of the threshold for when the digits merged together.

We know that the primary activity of our brain is to form internal models of our immediate environment and to mark each experience with one of three potent characters. The experience is either food to eat, a predator to avoid, or a mate to procreate with. These essential characteristics are common to all sentient animals and are a chief reason our limbic system remains in close contact with higher brain functions only recently acquired through evolution. An interesting feature of the limbic system is that, once some event is identified as a threat to our survival, it sets into motion a process that releases epinephrine, which can alter the permeability of nerve synapses. This allows them to more easily release neurotransmitters across the synaptic cleft and, in essence, causing the neurons to fire more readily and more rapidly under fight-or-flight conditions. Epinephrine, also known as adrenalin, causes an increase of heart rate, vasodilation, pupil dilation, and elevation of blood sugar levels. That means that when a person is highly stimulated (fear, anger etc.), extra amounts of epinephrine are released

An athlete experiences one form of psychological time in performing a skill while the audience experiences another, but both are embedded in an external world in which specific laws of nature exist in a larger time domain in which no two clocks measure the same passage of time

into the bloodstream to prepare the body for dangerous and extreme situations by increasing nutrient supply to key tissues.

Neural signals arising from any stressful situation activate the amygdala, which subsequently processes the information and activates the hypothalamus. Afterwards, the hypothalamus sends sympathetic discharges to the adrenal gland and facilitates the release of adrenalin into the blood. This in turn activates various autonomic responses to trigger the fight-or-flight response. This response happens physiologically faster than a person's consciousness is aware of them, which is a beneficial survival advantage because there is no thinking involved in order to take action. You may not even be aware that a circumstance is dangerous until after it has passed. During a dangerous situation, the mind must make a series of decisions about what to do: commonly described as Assess; Approach or Avoid; Fight-or-Flight; Freeze; or Collapse. During the Freeze and Collapse stages, the decision is to become immobile so that the predator might not actually see you, which is a characteristic behaviour we can see in deer, rabbits and squirrels all the time. In the Freeze response, everything freezes – time, thinking and all connection between the limbic system and cortex. The victim cannot respond to many of its sensory stimuli and may not be able to remember details of the attack. But you do not have to be threatened to trigger this dilation and

Muybridge 'The Horse in Motion' sequence of still images showing that all four hooves of a horse are off the ground at some point in a gallop.

slowing down of time. Your other emotional states can also be a strong influencer of how you experience the passage of time. Emotional events do indeed have the ability to speed up or slow down our subjective experience of time.

Another way of looking at this is what might be called the Cinematographic Principle. It is a common trick to speed up the frame rate of a movie camera to many times its normal 26 frames/s. When the video is then played back at normal speed, the motion will appear slowed down. It is not hard to imagine that during a time of stress, the brain records more Nows than it normally would so that it creates the psychological time in which to perform quick action. In retrospect when we study those memories, we have a sense of time slowing down during their playback. Mattis Appelqvist Dalton and his colleagues at Brown University have reviewed all of the various ways in which psychological time is both speeded up and slowed down. One of these ideas is called the Clock Model. '*In clock models, an internal clock emits pulses that are counted by an accumulator and then stored in memory for comparison to some baseline. When arousal is high, more pulses are emitted and accumulated, and thus durations are judged longer.*'

For the curious compression of time as you grow older, it seems that this also may have to do with how the brain stores memories and recalls them. This time, the memory may have nothing to do with the intense emotions spawned by triggering the limbic system. When an event is experienced for the very first time, the brain actively records it because it may have some kind of survival value later on. However, if the experience is not novel, other considerations may come into play and fewer of these events are remembered. The result is that children accumulate more memories of childhood summers than older adults do because so much of what they experience is novel.

WHEN RELATIVES COME TO VISIT

Time is relative.

Our subjective sense of time and its passing is an amazing organization of events and recollections of passing Nows, but they all take place within the larger world of the physical universe to which we belong and are embedded. Evolution has equipped us with highly specific brain circuitry that makes us keep track of what is going on around us. Events are ordered by cause and effect and our brains pay special attention to sequences of events – let's call them stories – that help us predict what may happen next. So, our understanding of time must move into the larger world with all its details both fathomable and unfathomable.

You would be very surprised to know that, although the human experience is rooted in the perception of Now, there is no such equivalent feature or phenomenon in the world outside the brain. This is identical to the fact that there is no such thing as colour in the electromagnetic spectrum because wavelengths of light are not tagged that way. Instead, colour is what brain researchers call a quale; a property of a perceived object created by the brain. Qualia are purely a figment of how our brains interpret sensory information in our internal, subjective, world. Similarly, there is no such thing as Now in the external world. This, like colour, may simply be a quale created by our brains to tag sensory information and improve survival. Visible light is colourless, but water and vegetation are 'blue' or 'green' and it is a useful distinction for the brain to discriminate between these two kinds of objects as quickly as possible. Colour tagging is one way to do this. Earth rotates about its axis and orbits our sun and it does so without any indication that Now is different from yesterday or tomorrow as the sliding Now moves ever-forward in time.

The failure to find Now in the world outside our brain is also reflected in the way we set up our theories about the physical world. All of our physical theories are expressed in mathematical equations using a symbolic language. That is because all of our measurements are based on numbers, and mathematics is the 'language' for ordering numbers in logical ways. The symbols appearing in our equations are called 'variables' because we can elect to change their values in the equations and then calculate a new number or prediction. We can then go out into the world and see if we can measure the same value to validate our theory. An example of this is the ballistic equation. It describes the vertical height of a

Complex spiral motion of a few charged particles in Earth's magnetic field. The tracks represent the worldlines of the particles from the reference frame of Earth's worldline frozen at a particular moment.

particle, z, starting from a height, h, launched with a speed of v_0 and being decelerated by the force of gravity whose value (on Earth it equals 9.8 m/sec/sec) is g.

A common variable we use to describe time is the symbol 't'. We write down equations to describe the flight of a baseball or a rocket, but these equations do not tell us where we are in the time variable 't' that we use to describe these systems. We can make this ballistic equation even more accurate by taking into account the detailed shape of the baseball, prevailing wind direction, and many other factors. We end up with a predictive equation that will very accurately tell us the height of the baseball above the ground to millimetre accuracy if we wish. Here is the problem. If I choose my Now to be the moment that the baseball was struck by the bat it is represented by t = 0. But if the baseball was struck in the past, then for the outfielder waiting to catch the ball, Now would just as easily be the point at t = 5.0 seconds after the bat's impact with the ball. This shows that our measure of time is rather arbitrary as to what absolute name we should give it, such as 14:35:06.6 on 3 March 2024. Perhaps this equation was meant to be the height of the ball struck by Babe Ruth in the Philadelphia Athletics vs New York Yankees game on 13 April 1921 at 15:30 EST. In other equations, why is the current Now time in the universe 13.8 billion years after the Big Bang and not 8.567 or 29.546? For that matter, why is my current Now located at the calendar date 26 April 2024 and not 23 July 1152 BCE or 23 November 3875 CE? The best theories we have about our physical world are completely

silent about how to pick out one moment, Now, from an infinitude of other moments.

Einstein's theory of special relativity published in 1905 at least gives us an important clue about time that can even be deduced from experiments in the laboratory and elsewhere. It is the first new insight about time we have had in thousands of years. We think we can organize our Nows in a sequence and using a clock, relate them to the Nows experienced by other people or even past historical events. The problem is that there isn't just one master time counted out by Mother Nature; in fact, there are an infinite number of different times. Every individual carries their own Now with them, but there is actually no universal clock to synchronize my Nows with your Nows or anyone else's. Let's see how this non-synchronized, multiple-time situation comes about.

Imagine a person, Emily, in a train car and another person, Stan, standing on the station platform. Emily has an instrument that can emit a single pulse of light directed at a mirror on the ceiling directly above the light source. The light pulse travels three metres up and three metres down for a total travel of six metres. Emily calculates the up-and-down light travel time (light speed = 3×10^8 m/s) for each pulse and gets exactly $Te = 6$ metres/(3×10^8 m/s) or

0.00000002 seconds. She enters this time in her notebook as 20 nanoseconds (20 ns). Meanwhile, Stan is not moving with the train as it speeds by at 100 kph. But he does glimpse Emily's experiment. Stan sees the pulse from Emily's clock travel along the hypotenuse of a right-angled triangle because the train is moving at $V = 100$ kph so that by the time the pulse has gone vertically up and back in Emily's frame, from Stan's point of view it has also moved sideways along the direction that the train is travelling.

Einstein's fundamental postulate in special relativity is that the speed of light, c, is an absolute constant. So, what is the relationship between these two times? Newton would have said that both clocks measure exactly the same time: Te = Ts. Einstein says not a chance, and he has a formula to prove it based on the geometry of the light pulse travel times and the constancy of the speed of light we just described.

Emily will measure the exactly vertical round-trip time to be Te = 20 nanoseconds, but with the train travelling at V = 100 kph (28m/s) and c = 300,000,000 m/s, Stan will see the light pulse take the slightly longer hypotenuse distance so Ts = 1.0000000000000044Te. If the train were magically moving at 50 per cent of the speed of light (150 million metres/sec or V = 0.5c) the time dilation becomes Ts = 1.154Ts so that Stan

A person on the train platform sees the pulse of light travel sideways along the black line from the floor to the mirror on the ceiling. A passenger inside the moving train sees the light pulse travel exactly vertically from the floor to the mirror on the ceiling and back to the floor because there is no sideways motion of the train in the passenger's reference frame. Because the speed of light is constant, the light travels the different paths by taking longer in the platform's frame of reference than for the passenger's.

would measure Ts = 23 nanoseconds. An observer moving with a clock in their sequence of Nows will see time pass more slowly than anyone else who is watching the moving clock. If there were a thousand different people, each moving at different speeds with respect to Emily's clock, they would be experiencing different Ts times; and none of them would be identical to each other's. In relativity, we call Emily's clock the proper clock, and the time it measures Te as the proper time. This sounds completely bonkers but this time dilation effect has been put to the test many times. It really happens even to dumb elementary particles that have no clocks.

In 1971, physicist Joseph Hafele and astronomer Richard Keating took four atomic clocks on commercial jets. Two clocks travelled around the world going eastward and two going westward. When they brought their clocks back to the US Naval Observatory and compared their times to the atomic clock in the lab that had not been travelling, the times were very different. The eastward clocks had lost 59 nanoseconds while the westward clocks had gained 273 nanoseconds. These matched the relativity predictions almost exactly and well within the measurement errors. So, the bottom line is that time dilation is a very real effect that causes the time measured by moving clocks to be different than stationary

ones. It is not an effect that has anything to do with the clock mechanism itself somehow being affected by the movement. Even the natural decay times of elementary particles are affected. Here are some other examples on the human scale.

When astronauts orbit Earth in the International Space Station, travelling at 8 km/s, the Doppler factor $(1-(V/c)^2)^{1/2}$ becomes 0.9999999996, and return to the ground from a six-month (1.5×10^7 sec) visit they will be about Ts = (1.5×10^7) x (1-0.9999999996) or 0.006 seconds younger than people that remain on the ground. In the far future when interstellar travel may be possible, astronauts taking a trip to Alpha Centauri some 4.3 light years from Earth may travel at a speed of V = 25 per cent the speed of light for a time dilation factor of 1.03. By Einstein's formula, their round-trip journey measured by their clocks would take about Te – 34.4 years but back on Earth our clocks would show Ts = 35.4 years had elapsed for the round trip. Things get even worse for trips to stars 100 light years away if the astronauts travelled at 90 per cent the speed of light. For those astronauts only Te = 111 years would elapse in proper time, but back on Earth with a time dilation factor at this speed of 2.29, about Ts = 254 years would have elapsed. This staggering mismatch between the two clocks would mean that instead of returning to Earth after a round-trip of Te = 222 years, it would actually arrive over Ts = 500 years in Earth's future. And now for the biggest thrill of all. Suppose you were riding on a photon of light at exactly the speed of light. Your time dilation factor would be infinite so that any trip you took in space whether to the moon or to the most distant galaxy would be instantaneous by your proper time clock, but folks back on Earth would see seconds or even billions of years elapse.

These examples, and tests, point out in stark detail how time is much more complicated than a cosmos having only one master clock that works for everyone in the universe, which is what Newton proposed back in the late-1600s. But Newton's mathematics only applied to situations that involved very slow-moving objects in space such as rockets or planets. Once you approach the cosmic speed limit of the speed of light, everything changes. Only Einstein's theory of relativity will give you accurate predictions of what will happen. Proper time (Te) is the only time that reigns supreme in this high-speed world. No matter where you are in the universe, how fast you are moving or accelerating, this proper time is your local measure of physical time with which you gauge events in your vicinity. All physical experiments will give the same results if they are expressed in terms of proper time and

use mass and length measures that are also travelling with the observer. Everyone else looking at the moving clock and moving length and mass measurements will see different values. Relativity and its equations let you work backwards from any observation to find what the experiment looks like in the moving proper frame, in this case Emily's train car.

Another important feature of relativity is that simultaneity, the condition in which two clocks keep the same time, does not really exist. It can never exist anywhere in our universe. No two clocks can ever be perfectly synchronized because time dilation will gradually bring the times recorded on the moving and stationary clocks out of synchrony. No matter how cleverly you try to set things up, you can never synchronize two different clocks to keep exactly the same time. Only *your* clock reads the correct (proper) time because it is travelling with you (on your wrist). No other clock can read the same time (be synchronized with yours to unlimited accuracy) unless it actually shares the same physical space on your wrist. But this isn't where the craziness ends in relativity. You see, it is not just a matter of how fast you are travelling, according to Einstein's theory of general relativity, it also matters where you are in space. General relativity says that time is inseparably wedded to space so that any calculation of *when* you are also has to include information about *where* you are. Even if you had two wrist watches strapped side by side on the same arm, they are not physically in the same place and so they will not be travelling along the exact same trajectory through space. Their times will slowly come out of synchrony.

THE MIND OF A LAZY DOG

Black hole with orbiting accretion disk. The geometry of 3D space is distorted by the presence of the black hole so that 'straight lines' are no longer straight but curved. Light rays from the back of an accretion disk are bent over the top and bottom of the black hole forming a circular halo. Meanwhile, time as measured by an astronaut falling into the black hole is greatly slowed down as viewed by a distant observer.

Albert Einstein published his major paper *The Electrodynamics of Moving Bodies* in 1905 in which he laid out the details of his theory of special relativity, but by then many of his conclusions had already been known for at least a decade. The major insight Einstein provided was his proposition that the speed of light was an absolute constant no matter how you measured it. His proofs and discussions were very mathematical, but they eventually caught the attention of his former geometry professor Hermann Minkowski at the Eidgenössische Technische Hochschule in Zurich. As it turns out, young Albert was not much of a student and would routinely skip Herr Minkowski's classes, but passed his maths tests using notes provided by his girlfriend. Minkowski's impression of Albert was that he was a marginal maths student and a 'lazy dog'. When he saw Einstein's paper published in the prestigious *Zeitschrift für Physik*, he was shocked. *'Imagine that! I would never have expected such a smart thing from that fellow.'*

Hermann's speciality was abstract geometry, but he quickly saw in Einstein's paper that Einstein had been groping for a geometric description of relativity without properly understanding the underlying geometrical nature of his own main ideas. In short order, Minkowski filled in the technical details and it is Minkowski's vision and language of relativity that persists today. Minkowski's work was at first not received very enthusiastically by Einstein, who dismissed Minkowski's translation of special relativity as '*superfluous learnedness*'. But Minkowski persisted in his revolutionary vision. Eventually, Einstein adopted many of Minkowski's methods in 1912 in the development of his new theory of general relativity. For Einstein, time and space were still separate physical concepts, but for Minkowski's geometry this separation could not work for special relativity. As he famously noted: '*The views of space and time which I wish to lay before you have sprung from the soil of experimental physics, and therein lies their strength. They are radical. Henceforth, space by itself, and time by itself, are doomed to fade away into mere shadows, and only a kind of union of the two will preserve an independent reality.*'

Spacetime is vast. It extends well beyond the Earth and solar system, encompassing the entire three-dimensional universe out to the farthest galaxy and beyond. Its indivisible time-like aspect also extends from the instant that the

universe flashed into existence, through the present moment, and on into the future. But what is spacetime, really? Although popularizers love to describe spacetime as some kind of rubber sheet that can be ironed flat, or rolled up into a ball, in fact it is none of these things. Spacetime can no more be felt than can a thought, or the landscape of a dream. It is not a place in which we live, but a condition in which we exist. Unlike thoughts and dreams, however, it appears to have a precise geometry and it is through this geometry that every action and event in the universe is defined. In some ways, spacetime is to the physical world what the term 'economics' is to money. You can no more divorce spacetime from the physical world than you can speak of money without invoking economics.

Just as Euclidean geometry can be drawn on a flat sheet of paper, the details of spacetime are recorded in the 'worldlines' that criss-cross its countenance and actually synthesize spacetime itself. As Minkowski also notes about this fact:

> *'The whole world appears resolved into such worldlines. And I should like to say beforehand that, according to my opinion, it would be possible for the physical laws to find their fullest expression as correlations of these worldlines.'*

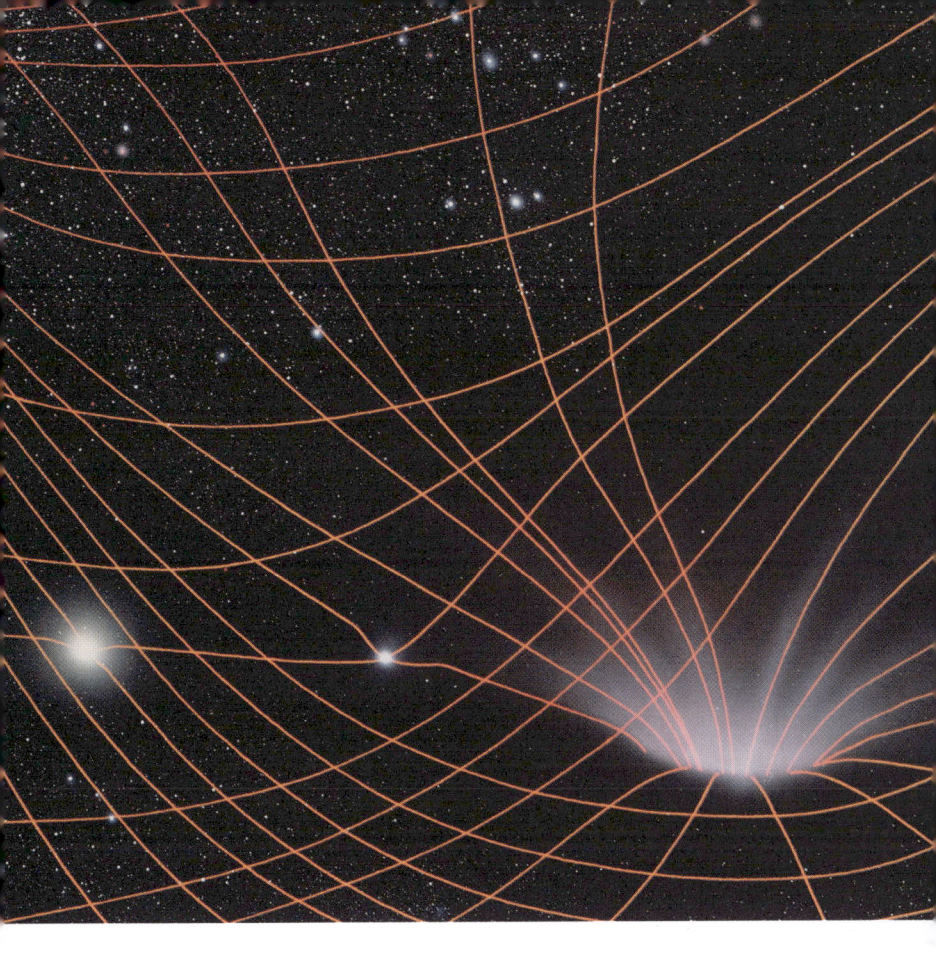

The shapes of the worldlines of particles define the geometry of spacetime as shown by this artist's rendition of coordinate lines drawn in the warped space of a black hole.

Physicist Lee Smolin, in his book *Three Roads to Quantum Gravity*, describes the essential foundation of relativity as the 'story' about processes and not the things-as-objects. '*A marble is not an inert thing, it is a process.... There are only relatively fast processes and relatively slow processes. And whether it is a short story or a long story, the only kind of explanation of a process that is truly adequate is a story.*' We cannot define an object, be it a human, a table, or an electron by merely describing its properties at one instant in time. We can only define an object in terms of a process consisting of innumerable events that create the story that defines it. In relativity, the history or 'story' of a process such as a football or a galaxy, consists of a series of events that are tied together by cause and effect to create the process that you see at any particular moment. These events include the interactions of one process with others that cumulatively create what you see as the history of the process at a particular moment. In relativity, we call this history of a process its worldline. What Einstein said is that only worldlines matter, because these are the only things we have access to. Even better than that, we are only able to see that part of a process that can be communicated to us by using light, which is the fastest signal we can ever use to transfer information.

The beauty and deep mystery of Einstein's relativity is that all of its explanations and predictions are launched from a world in which the three dimensions of space and the one dimension of time are not treated separately. Spacetime becomes the four-dimensional arena in which the worldline histories of every particle in the universe are traced out as sequences of events. The exact timing of each event on a worldline is guided by the proper time of the clock travelling along the same worldline or one that is very close by. Your worldline began at your moment of birth and will end at the moment of your death. For me, I was born on 23 November 1952 at 10:25 am (09:25 UT) in a hospital room in Karlskoga, Sweden located at latitude 59°18' 54' North and longitude 14° 30' 24' East. The radius of Earth is 6,378 km, so relative to the centre of Earth my location in spacetime is a set of four numbers in decimal form (1952.8992, 6,378 km, 59.315°,14.507°). After this event in spacetime, my worldline becomes very complex as I move around in space and grow older, and I don't know what its endpoint will be in time or space.

Spacetime is not some abstract mathematical arena. Einstein in 1915 published his second paper on his new theory of general relativity in which spacetime became another name for the gravitational field of the universe in which time

and space are the fundamental components. This is identical to how we describe light as an electromagnetic field in which its magnetic and electric aspects are the two components. Thinking of spacetime as a gravitational field led to a whole new series of predictions for new phenomena in nature such as black holes, Big Bang cosmology, gravitational waves and a peculiar effect called gravitational time dilation. It was no longer necessary for clocks to be in motion to show a dilation effect. They just had to be in locations in space where the local gravitational field of a planet or a star was different. One of the most exciting proofs of this phenomenon, and that gravity causes spacetime to be curved, came from the Pound–Rebka experiment performed by physicists Robert Pound and Glen Rebka at Harvard University in 1959.

Imagine two experimenters, Thelma and Louise, where Thelma is standing on the ground and Louise is on top of a tall step ladder. Thelma defines a time interval by sending exactly 10 cycles of yellow light to Louise, who then waits until her equipment has registered all 10 beats of the light. According to general relativity the light signal Louise receives has lost some energy climbing up the Earth's gravity well to the top of the ladder and has been shifted to a lower frequency. Louise has to wait longer than it took Thelma to send all 10 beats of the light signal in the first place. The reason that this experiment contradicts special relativity is because, first of all, both Thelma and Louise are at rest relative to each other and moving with exactly zero velocity with respect to each other. This means they should be sharing exactly the same 'proper' reference frame in special relativity. But this necessarily means that a single, flat spacetime patch should be exactly sufficient upon which to describe both their worldlines, and those of the wavelets in the light ray they exchanged. Their worldlines should form the opposite sides of a rectangle with perpendicular sides, and in which the light ray forms the diagonal. By a simple geometric construction, however, we see that this cannot occur for Thelma and Louise because the light wavelets take longer to receive than to send. This means that the spacetime geometry between Thelma and Louise has to be distorted in some way. The rectangle becomes a distorted trapezoid with unequal sides.

The Pound–Rebka results can only be accommodated if the spacetime patch, which the worldlines of Thelma, Louise and the light rays share, is not flat, thereby showing that the gravitational 'redshift' demands spacetime curvature. This curvature of spacetime by gravity was also shown in

the famous total solar eclipse observations on 29 May 1919 by Sir Arthur Eddington. General relativity predicted that, close to the sun, the images of distant stars should be shifted from their normal positions in the sky when the sun was not there. The amount of this shift was found to be almost exactly what Einstein predicted. Recently, a precision atomic clock was created that reveals just how weird the local 'flow' of time can be.

Physicist Tobias Bothwell and his colleagues at the Joint Institute for Laboratory Astrophysics (JILA) in Boulder, Colorado, separated thousands of strontium atoms into pancake-shaped blobs of about 30 atoms each and put them into a vertical stack just one millimetre high. General relativity would predict that even across this miniscule one millimetre span of heights, Earth's gravity will cause the frequency of oscillations in each group to shift by a different amount. The amount of this delay similar to the Pound–Rebka effect would only amount to a time difference of about one part in 10^{19}. Previous measurements had observed the redshift over larger scales by comparing separate clocks, but the JILA team measured it within a single clock.

You can only synchronize clocks by knowing their detailed history of movement in Earth's gravitational field. The same synchronization problem applies to any two clocks in the rest of the universe. Light signals take time to travel from point to point. You cannot know the exact path the light ray took to give you the start signal to synchronize two clocks. You can also not keep them ticking in synchrony because they are moving across space through different gravitational fields and this constantly changes their time dilation factors.

What these experiments say is that when we look closely at what we mean by space and time they are far more complicated than what our intuitions might suggest. In the hands of general relativity, which gives us phenomenally accurate predictions, space and time can be distorted or warped in the presence of anything that produces a gravitational field, such as matter or even raw energy like photons of light themselves. In essence, using a small collection of experiments and observations, our brains have fashioned a new model for space and time in the world beyond our brains. Our exploration of the nature of spacetime also leads us into a very peculiar realm of thinking even as we execute a very mundane action.

I take a walk to the store and can't help but feel I am moving through something that is more than the atmosphere that rushes by my face as I go. The air itself is contained

within the boundaries of the space through which I pass. If I were an astronaut in the vacuum of outer space, I would still have the sense that my motion was through a pre-existing, empty framework of three dimensions. Even if I were blind and confined to a wheelchair, I could still have the impression through muscular exertion that I was moving through space to get from my kitchen to my living room 'over there'. But what is space as a physical thing? Of all the phenomena, forces and particles we study, each is something concrete though generally invisible: a field; a wave; a particle. But space, itself, seems to be none of these.

Back in the early 1700s, Sir Isaac Newton proposed that space was an ineffable, eternal framework through which matter passed. It had an absolute and immutable nature. Its geometry pre-existed the matter that occupied it and was not the least bit affected by it. A clever set of experiments in the 20th century finally demonstrated rather conclusively that there is no pre-existing Newtonian space or geometry 'beneath' our physical world. There is no absolute framework of coordinates within which our world is embedded. What had happened was that Albert Einstein developed a new way of thinking about space that essentially denied its very existence!

Einstein's relativity revolution completely overturned our technical understanding of space and showed that the entire concept of dimensional space was something of a myth. In his famous quote he stressed that, '*We entirely shun the vague word "space" of which we must honestly acknowledge we cannot form the slightest conception. In the relativistic world we live in, space has no independent existence...[prior-geometry] is built on the a priori, Euclidean [space], the belief in which amounts to something like a superstition.*' So, what could possibly be a better way of thinking about space than the enormously compelling idea that each of us carries around in our brains, that space is some kind of stage upon which we move? To understand what Einstein was getting at, you have to completely do away with the idea that space 'is there' and we move upon it or through it. Instead, relativity is all about the geometry created by the histories (worldlines) of particles as they move through time. The only real 'thing' is the collection of events along each particle's history: the succession of Nows that make up your history. If enough particles are involved, the worldlines are so numerous and close together that they seem like a continuous space. But it is the properties of the events along each history that determine the overall

geometry of the whole shebang and the property we call 'dimension', not the other way around.

The wire-frame illustration opposite is an example where the wires (analogous to worldlines) are, which define the contours of a dimensional shape. There is nothing about the background (black) space that determines how they bend and curve. In fact, with a bit of mathematics you could specify everything you need to know about the surface of this shape and from the mathematics tell what the shape is, and how many dimensions are required to specify it!

Princeton University physicist Robert Dicke expressed it this way, '*The collision between two particles can be used as a definition of a point in [space].... If particles were present in large numbers...collisions could be so numerous as to define an almost continuous trajectory.... The empty background of space, of which one's knowledge is only subjective, imposes no dynamical conditions on matter.*'

What this means is that so long as a point in space is not occupied by some physical event such as the interaction point of a photon and an electron, it has no effect on a physical process (a worldline) and is not even observable. It is a mathematical 'ghost' that has no effect on matter at all. The interstitial space between the events is simply not

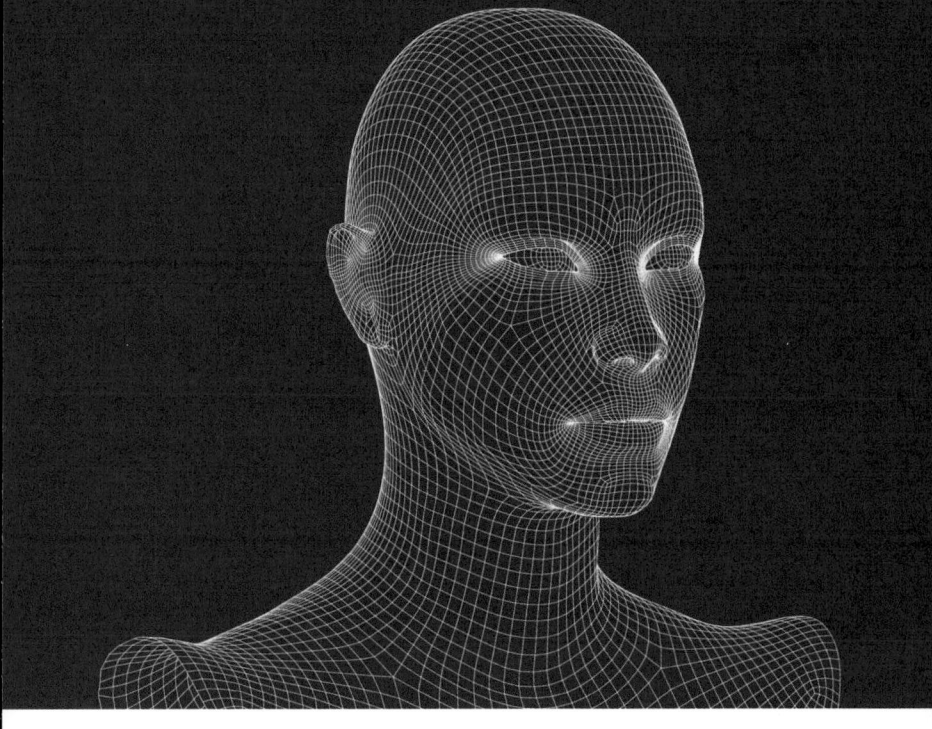

Wire-frame head representing how curved worldlines can combine to define a geometry for space even though space itself has no dimensionality or geometry of itself.

there so far as the physical world based upon worldlines is concerned. It is not detectable even by the most sophisticated technology, or any inventions to come. It does not even supply something as basic as the 'dimension' for the physical world!

We should also be mindful of another comment by Einstein that '*..time and space are modes by which we think and not conditions in which we live*.' They are free creations of the human mind, to use one of Einstein's own expressions. Like dew on a spider web, individual and numerous events along a worldline define the worldline's shape, yet like the spider web, this web can be thought of as embedded in a larger domain of mathematically possible events that could represent physical events...but don't. The distinction between these two kinds of points is what Einstein's revolutionary idea of relativity provided physicists, and is the mainstay of all successful physical theories since the 1920s. Without it, your GPS-enabled mobile phones would not work!

So, what are these events? Simply put, according to physicist Lee Smolin, they are exchanges of information, which are also the interaction points between one particle's worldline and another particle's worldline. If you think at the atomic level, each time a particle of light interacts with (collides or is emitted by) an electron it generates an event. These events are so numerous the electron's worldline looks like a continuous line with no gaps between the events. So the shape of one worldline, what we call its history, is a product of innumerable interactions over time with the worldlines of all other objects (photons etc.) to which it can be in cause-and-effect contact.

So what does all of this talk about worldlines and fictitious events have to do with time? As it turns out, thanks to Einstein's theory of general relativity, we have an explanation for the whole universe that includes gravity, space, time and worldlines. Worldlines are four-dimensional objects. A collection of these worldlines, which we can liken to a story or a history, and their various interlinkages (interactions) defines a geometrical framework that has a shape. In relativity, the collection of worldlines creates a four-dimensional volume that consists of all physical events accessible to the worldlines. In many ways, it resembles a network-like structure.

What we also begin to see is that, when we think of spacetime as an ensemble of worldlines connecting physical events, the dimension of time becomes hidden as just another link between events that is not unlike the way in which events are linked in space. But at the level of individual worldlines and events, we still do not see an explanation for why we persistently experience time as a very separate phenomenon from space and furthermore why nature and the universe seem to also go along with this separation.

THE CHOICE IS CHICKEN, STEAK OR FISH

Artistic rendering of the Big Bang. A single hydrogen atom will follow a tortuous worldline as it moves into and out of many different systems of matter, ending up in the glass of water you just drank on Earth today.

Imagine the worldline of a single hydrogen atom created in the Big Bang. It starts out in its proper time frame at the Big Bang and wanders through intergalactic space until it is finally incorporated into the matter of the Milky Way galaxy where it takes up residence in one of the many interstellar gas clouds. After a long time the gas cloud collapses to form our sun and solar system but the hydrogen atom finds itself as a part of Earth and then through many billions of years of cycling between the atmosphere, plant and animal tissues, it ends up a few months ago as one atom in the glass of water you drank. From there your internal organs and cellular mechanisms use it in a cell division step that produces a DNA strand. This takes us all the way to the present moment. But Einstein's relativity says that worldlines only end when the history terminates and the object vanishes. So far as we know, the history of this particular hydrogen atom continues to be a part of Earth in the future until our sun becomes a red giant star and incinerates Earth, releasing the hydrogen atom back into interstellar space. After that, the only way that it terminates is in the far future when the universe itself ceases to exist.

If we consider all the other objects that came into existence at the Big Bang, there are over 10^{88} of these worldlines making up the histories of every particle in the visible universe, together with many more that came into existence in supernova explosions, cosmic ray interactions and other fission/fusion events across the universe. Together, all of these worldlines define a four-dimensional volume in which each point is threaded by the worldline of some particle. This works well if you think in terms of all the past events. You can imagine having a cube whose left-hand face is inscribed with the words Big Bang, and the right-hand face with the word Now. The worldline started at Event A (the Big Bang) and came to an end at Event B (Now).

When we look at the worldlines of galaxies across the universe, we see that they all point back to a specific era in time about 13.8 billion years ago in each and every particle's proper time frame. Every object in the universe identifies this same event as the origin of its own worldline, except for those few that were created later on in supernovae or by collisions along the way to the present. But when we get to the present Now, it looks like all the worldlines come to a sudden end. But in fact they don't. According to the rules for how we are to create worldlines in relativity, for every Now

moment, the particle would have to be utterly disintegrated, but then mysteriously brought back into existence in the next instant. We see that the wall representing Now is steadily moving to the right into the future, as one Now is replaced by the next one. The catch is that in both special and general relativity, worldlines can be mapped into the future using the laws of physics, principles of conservation of energy and momentum, and cause and effect. Histories of many objects in the universe can be mathematically projected into the future using their past worldlines as a guide. We do this all the time when we forecast tomorrow's weather, or the detailed circumstances of a total solar eclipse on 17 September 2992 over North America. We even do it when we are driving a car on the road or walking. Because we can make these predictions, there are at least two ways we can interpret the result.

In one view, known as eternalism, past, present and future events *all exist together* in what is called the 'block universe'. The worldlines of every particle that emerged from the Big

Stroboscopic montage of an athlete. In the block universe, past, present and future coexist all at once.

Bang exist in their entirety from their instant of creation to their instant of vanishment in the far future, along with all the interactions and specific events that make up their worldlines. Extending our previous discussion, we see that Now is just one slice through this block of spacetime. The entire block exists all at once because time is an integral part of its description and so the block does not itself change in time; it simply exists all at once. In this block universe model, our inability to perceive future events is simply an illusion. As physicist Hermann Weyl noted, *'The objective world simply IS. It does not happen. Only to the gaze of my consciousness crawling upward along the worldline of my body does a section of the world come to life as a fleeting image of space which continuously changes in time.'* This version of the world is a delight to time travel enthusiasts because it says there are in fact past events that still exist 'out there' that we could revisit if we had the right technology. Even more confounding is that, if the worldlines of every atom in your body, every electron in its orbit, every neurotransmitter on its way to a synapse, are already defined, there can be no free will. In the block universe, nature gives you time travel with one hand, but takes away your free will with the other. Free will is a complete illusion created by the brain just like colour. There is no colour 'red' mixed in with the wavelengths of electromagnetic radiation, just as there is no free will mixed in with the bundle of worldlines we inhabit and the nature of the next events. But do the events taking place in the year 2745 CE exist with the same solidity as the events of today? Eternalism says yes, but your hunch and common sense says no. You are in good company because although most physicists may use eternalism to predict tomorrow's weather, they don't fully subscribe to the idea that tomorrow exists in the same way that the pyramids of Egypt do, or that the past is really 'out there' to be revisited in a time machine.

A contrasting idea to eternalism is presentism in which the past and future do not exist but only the present has any physicality to it. Presentism is in far greater accord with our sense that the past is ended, the present is immediate, and the future is open and affords us some free will over which specific futures we may choose. It is also consistent with how our internal sense of time operates as we saw in the previous chapters. Presentism declares that the past appears to exist, but only as a collection of latent information resembling a hologram encoded in three-dimensional physical records (rock carvings, photographs, video tapes, etc.) strewn around

our spatial environment and still accessible in the present Now moment. The universe is three-dimensional, but time as a continuum of events does not really exist at all. Only Now exists in the elusive present.

The past is a hologram – the future is a cloud of uncertainty. I wrote this sentence at 5:03 pm EST on 5 May 2024 from my desk located at the geographic coordinates +39° 00' 50.23' North and 77° 5' 16.72' West. This moment of time is in my present, what my brain is experiencing as Now, but at the same time in your Now on your worldline it is in your historical past. If you were clever, you could visit my location in your Now and search for clues in the objects at that location that could help you reconstruct my surroundings and perhaps even that photo I took of myself writing the period at the end of this sentence. But from your vantage point in the future, and on your worldline, I am a complete phantom, even though from my perspective I am very much alive. Also, by the time you reach the end of this essay, I will have changed your brain state. The information I just transmitted to you in this paragraph from my worldline to yours is now being incorporated into some of the neurons in your brain for your future access. But this process is not symmetrical. I can change your brain state, but you cannot reach back in time and change mine as I finish writing this sentence... right...*Now.*

Presentism asserts that the grand scope of time we think is 'out there' along our worldline is actually just a set of information coded in our 3D world. We can, however, assemble a clearer view of the past stored in the present Now by assembling all the records encoded in three-dimensional space about a specific event or object. This is like finding all the horcruxes to reassemble Voldemort, or a detective recreating a crime scene from scattered clues at the location: a gun, a letter, drops of blood in the carpet. Entropy, however, will degrade these records depending on their distance from the current moment. Long-ago records will contain less information having suffered through many ensuing years of entropic increase. Meanwhile, recent records have suffered through fewer years of entropic increase and will be fresher and less eroded. Does this make sense? Many physicists have trouble with the idea of presentism because it seems to invalidate the basic principles of relativity. How can you have a singular Now for the universe when you cannot synchronize clocks to define a single common spatial surface? Presentism was popular decades ago but has now fallen out of favour because it is

A hologram captures the 3D details of an object by coding the light information in a 2D surface or piece of glass.

incompatible with relativity.

An interesting combination of presentism and eternalism was offered in 2009 by physicist George Ellis at the University of Cape Town in his crystallizing block universe theory. What we call the present is the boundary between two regimes where the indeterminacy of the future is crystallizing into the certainty of the past. Humans ride this crystallization boundary through their perception of their individual Nows and the features of our local world run by a single clock on the wall. This has some interesting consequences for time travel. It basically puts an end to this idea. You can no more visit your past in a time machine than you can step inside a hologram.

So is our menu choice steak, chicken or fish? Is it eternalism, presentism or crystallization? All of these discussions provide us with a way of thinking about time on the human and cosmic scales but do not bring us any closer to the essential mystery of why spacetime contains three dimensions that are purely space-like along with exactly one dimension that is purely time-like in which spatial states seem to advance and change. Even though relativity shows us that time is plastic, it does not explain why it is a dimension in spacetime singled out for special attention. We can follow this unique dimension all the way back to the Big Bang itself and find it nearly impossible to discern exactly how it emerged...nevertheless it did!

MEMENTO MORI

A portion of the King List at Abydos showing the names of a few of more than 100 pharaohs in the sequence in which they reigned. Records such as this that are embedded in 3D space, can be used as a 'horcrux' to reconstitute the progression of these pharaohs along the time axis when combined with the static records of the other pharaohs in the list that reinforce their ordering via causality. Some names have been obliterated by the accumulation of erosion (entropy).

What we have learned so far helps us understand how we internally create the sense of time we have. Observations of the rest of the universe help us understand how things operate within time to move, to evolve, to change. We have described the organizing principle of the worldline and how events are organized along it using local proper time marked by the ticking of a combination of other worldlines that we call a clock. We have even attempted to understand the big picture by exploring eternalism, presentism and the hybrid notion of the crystallization of the past from future possibilities. Are any of these explanations for why the organization of things in space along a fourth dimension called time is really getting to the heart of why time exists in the first place? Thanks to Einstein's relativity theory, we now see that time and space are so intimately intertwined that, like the Gordian Knot, they cannot be disentangled without doing violence to spacetime itself.

The previous ideas rely on the idea that the things we see and experience at one moment are related to other things by being organized in a framework of cause and effect, which is called causality. Every worldline is just such a series of events related to each other by causality. For example, your family tree is a collection of individuals represented by nodes and links, but the organization of these individuals is not explicitly in time. Each node is a given individual connected by a link to their direct ancestor (parent, grandparent etc.). Other links show the individual's direct descendants (child, grandchild, etc.). The only coded information is causality (parent caused child caused grandchild). When we create a pedigree chart, time is often not even mentioned at all. In many instances, we know the names of an individual but not their dates of birth or death. Is there anything we might learn from this kind of representation that might help us towards a deeper understanding of time? According to British physicist Julian Barbour at the University of Cambridge, there most certainly is.

Barbour's idea is that we never experience time directly, no matter how we try to do so. The past is never anything more than what we can deduce from a series of present-day records. Instead, like the colour of specific wavelengths of visible light, it is a quality we infer from specific information we have gathered in the here and now. Never mind that the gathering and deduction process occurs within the

neuronal discharges of our brain over a period of time, making this idea sound like an exercise in circular reasoning! In physics we do have important equations that have no variable corresponding to time, t. These equations can be thought of as representing a static landscape. One of these equations is called the Wheeler–DeWitt equation and its landscape, called superspace, consists of all possible ways in which the three-dimensional space of our universe can be configured.

Superspace is a mathematical construct. It may not actually exist except in the human mind. Each point in superspace is an object that represents the complete geometric content of our 3D universe. Superspace is an infinite and timeless 'place' that contains all of the possible 3D geometries for our universe. Our specific universe is coded by a path through superspace whose probability is believed to be the most likely of all possible paths connecting the individual points. Some physicists believe our universe is defined by the path that has the greatest likelihood, and so it contains many individual points that in turn have

An example of paths in superspace connection of the most probable points, each representing a 3D snapshot of our universe.

the greatest probability as determined by causality. This means the path follows the ridgeline of many high peaks in superspace. Time, however, is not an element of this construct. But our universe represents a path through this landscape of possibilities. Along the path are a succession of shapes for 3D space, and configurations of matter down to the smallest quark and electron.

The selection of which state comes next is entirely based on the principle that our universe represents a sequence of timeless states that make the likelihood of our universe existing as high as possible. According to Barbour, the states with the highest probability are those that also include information about which state or states immediately preceded them. Not surprisingly, Barbour calls this information a time capsule. For example, if you visit the Giza Plateau in Egypt, you will see pyramids. These pyramids are time capsules of information in space that contain information of the existence of other states, ancient Egyptians, that came before them according to causality. If a state has many such time capsules, it can make other states increase their likelihood too, because they form a correlated series of states of increasingly higher probability. This is like the way that Google ranks websites according to how many other websites link to them. The highest-ranking websites are where the preferred search results are found. By consistently comparing time capsules we can build up vast histories of the world and universe along the path without any reference to time passing. Is this a crazy idea? Not necessarily. We have our own brains to thank for a dramatic example of this process at work.

We saw earlier how the brain knits together sensory impressions spanning about 100 milliseconds into an impression of Now. Most of this integration is done subconsciously and because of the different time lags between sensory organs and this point in the neural processing, a strict time-ordering is not always obeyed. A physical example of this can be found when we watch a person bounce a ball more than 40 metres (130 feet) away. Sound travels slower than light so that the audio and visual information is pulled out of synchrony at each bounce. We see the ball strike the ground before we hear it. A similar kind of asynchrony is also present based on the arrival of synaptic stimuli into the prefrontal cortex. What is interesting is that this window of about 100 milliseconds is not fixed. The processing is done via neural discharges and collections of these can have their own time lags too. The result is that some trace of the information from a previous Now lingers into the processing of the current Now. This lingering of past information into the present is believed to be the reason we have a sense of continuity in time. Instead of our Now being somehow chunked into fixed segments of time, it actually has a width that can last several seconds. This larger window has been called the 'specious present' by the philosopher William James, who turned the concept into an actual school of study. But in the end, all of this previous information is stored like a time capsule 'memory' among the three-dimensional spatial neuronal synapses of a brain circuit. This is the same idea that Barbour proposes for the world at large: *'Yesterday seems to come before today because today contains records – memories – of yesterday. The Nows with the most "tickets" have the best chance of occurring.'* He envisions a massive and almost incomprehensible landscape in which all possible Nows exist. The vast majority do not contain any records or time capsules of other states and so do not increase the likelihood that a particular universe path through this landscape will occur. So how do these individual states along a path give rise to the very obvious sense of movement? Here, Barbour joins Hermann Weyl in suggesting that our sense of time passing is an illusion (see page 86). The mystery of quantum mechanics is that it can indeed provide movement without actual motion as we will see in the next chapter. There is another feature we also need to add to this description: entropy.

Recall back on page 8, that Saint Augustine famously said, '*What then is time? If no one asks me, I know what it is. If I wish to explain it to him who asks, I do not know.*' To describe what physicists mean by entropy is a lot like Saint Augustine's complaint but luckily we know a lot more about it. Entropy is often used as a means to gauge how much disorder there is in a system, but that definition misses the point. Entropy is a measure of how many alternate states are available to a particle or system. For example, in my kitchen I have a floor covered with ten equal parquet squares. I toss a ball on to the floor and it comes to rest in one of the squares. Because there are ten squares, there are ten possible places that the ball could end up. Now imagine my dining room with 10,000 parquet squares. Again I toss the ball into the room and it comes to rest on one of the squares, but now there were 10,000 possible places were it could have landed. Because entropy is a measure of the number of possible alternate states for the ball, the entropy of the kitchen is lower than the dining room. Because for gas molecules the number of available states increases with temperature, we say that

hot gases have higher entropy than colder gases. Physicist Ludwig Boltzmann actually worked out a simple formula for calculating entropy as

$$S = k_B \ln W$$

where k_B is a constant (called Boltzmann's Constant), W is the number of possible states available to the system, and ln W is the natural logarithm of W. In our example, for the kitchen, W=10 and so we have $S = k_B \ln (10) = 2.3k_B$, while for the dining room, W=10,000 and so we have $S = k_B \ln (10000) = 9.2k_B$.

An important consequence of counting states and entropy is that the entropy of a system can never decrease. This is called the second law of thermodynamics. The important caveat is that it must be a closed system in which it is not gaining or losing energy from its environment. For example, if you let a number of gas atoms into a box and let them rearrange themselves, they will tend to settle into a most probable state where they are spread out and uniformly occupy the volume. You will never find them suddenly surging into only a smaller number of states with lower entropy. However, if the system is not closed, it could be chilled by an external refrigerator, in which case the entropy of the chilled gas could be lowered. When atoms or any system reaches a state where it occupies all of the available states, we say that this system is in thermodynamic equilibrium and thereafter does not change further – its entropy is maximized. The nice thing about this process is that we can force a system to be out of thermodynamic equilibrium and use this to do useful work. Living systems are only in thermodynamic equilibrium and 'maximum entropy' at the time of death, but by ingesting food at lower entropy, they can persist for long periods of time out of thermodynamic equilibrium.

Our universe is believed to be a closed system. Although it was much hotter in the past during the Big Bang, because of the relentless expansion of the universe there was less space for matter and radiation to occupy. Compared to today's colder and larger universe, the entropy of the early universe should actually be dramatically lower than it is today. This is because there is one ingredient we have not included: black holes. The entropy of a black hole can be calculated and it depends on its surface area. The single supermassive black hole at the centre of the Milky Way has an entropy equal to everything that came out of the Big Bang that now fills our entire visible universe. So even though it looks like entropy has decreased among the cooling cosmic gas and radiation

According to a holographic model of black hole event horizons, they record information that can specify how the event horizon was formed, and possibly events occurring inside the event horizon itself.

in our present universe, the formation of black holes has ensured that the Big Bang had a much lower entropy near the Big Bang than the universe today.

Another indicator of entropy in action is the production of waste heat. Whenever you assemble matter into an organized state you are doing work to arrest its decay into thermodynamic equilibrium. For example, you extract iron ore out of a mine. It used to be in a stable, random state but now you have expended work to extract it into an unusual state in which the iron atoms are no longer in equilibrium with the rock environment. But to do this extraction from a high entropy to a low entropy more-organized state, you have had to expend energy in your muscles and this produced heat as your body warmed up from the effort. Closer to home, a chlorophyl molecule in a plant takes in a single photon of sunlight in the visible spectrum and uses this energy to forge molecules for its internal systems such as cellulose. The chemistry generates several infrared-wavelength photons of waste heat at the end. So, plants use sunlight to lower their entropy against the relentless tendency for its chemistry to want to reach thermodynamic equilibrium in a higher-entropy state – called death. What does this all have to do with time?

In the cosmos, the direction of physical time passing, what we experience as Nows changing from the past to the future, is also the direction in which entropy is increasing. What this means is that if we write our name on a piece of paper we

are creating a low-entropy system frozen at that moment in time at the expense of slightly increasing the entropy of our local universe as we permanently rearrange atoms and expend some heat energy. But if we skip ahead a hundred years, that piece of paper will degrade as the entropy of the universe relentlessly increases. Meteors create organized structures called craters but do this by generating waste heat. The neurons in our brain store memories which are low-entropy organizations of matter, but generate waste heat during the formation process. So, whether it is in craters, monuments, time capsules or memories, traces of the past can be found in the present because entropy was lower in the past to form them. On a pavement you notice a cracked egg. Our assumption of low entropy in the past allows us to say that not long ago there must have been a more ordered, unbroken egg. The egg on the pavement is like a memory in your brain. It is a record of a prior event but *only* if we assume a low entropy condition in the past.

Part of the distinction we draw between effects and causes is that effects generally involve an increase in entropy. This assumption that the past has a lower entropy than the present going all the way back to the Big Bang means that the Big Bang had to be a state of lower entropy; but we cannot derive this unambiguously from observations and data so it is instead called the past hypothesis. When we create a history of some object, we can only select a sequence of states that obeys this hypothesis. This hugely limits the number of different stories we can create from the enormous range of possibilities. So, when we look at Barbour's huge stack of possible Nows and the many alternate stories they can be organized to tell, we can only consider those that conform to the past hypothesis, and where we can deduce cause and effect and low-entropy to high-entropy progressions from how these many Nows and their time capsules are organized.

On the human-scale, my time capsules consist of diaries, photos, video records, books and articles kept in my home, together with many copies of these kept by other people who know me, or have purchased my books around the world. There are even BBC video recordings of me at TV stations in London and church records of my birth in Karlskoga, Sweden. I have many such 'horcruxes' embedded like Barbour's time capsules in 3D space, which can be assembled via causality and entropy into a time-like sequence. Eventually, however, random fires and other mishaps will erase all of them from Barbour's ensemble and I will be no more.

THE THEORY THAT NO ONE UNDERSTANDS

A night-time view of the United States shows from left to right how random lights in the Midwest seem to organize themselves into regular shapes going eastwards. This may be akin to how time emerged in a phase transition from non-causal chaos in space to causal order along a new axis.

We have searched for the origin of time within our own brains but found merely a subjective tapestry of Nows held together by memories and expectations. We have also searched for the origin of time among the laws and theories that make up the world beyond our bodies. In each arena, we learned new things and the means for thinking about time as another property to explore like the surface of the moon or the distant stars. But there is still something missing. Although we know how to measure time with great accuracy, and we can forecast the next moments in a bewildering number of things in space, the deep nature of time still remains aloof. We can't even formulate a question about our world that does not involve time itself. This is because we seek more than just a collection of photographs; we look for the story behind them in virtually every experience we have of the physical world. The fact that entropy is increasing and that we have to accept the past hypothesis (there was lower entropy and disorder in the past than today) helps us organize the world into past and future, it still doesn't seem to explain the essence of the experience of time itself. So, is there any other arena where we might look for answers to time's origins? Having explored the world-in-the-large, and the world within the brain, we now delve deeply into the world of the atom and the quantum.

Before we take this next step into the quantum world we need to recognize how mathematics is used to help us visualize what is going on in domains not accessible to our senses or scale. Physicists since Newton rely a great deal on using equations to predict what an object does next, and sometimes those equations are vivid and match what we see with very high fidelity.

Consider the parabolic path of a projectile. Its trajectory is described perfectly by two equations; one for the vertical motion and one for the horizontal motion. Together they describe a parabola frozen in time, which is the observed path taken by the projectile in space. Far more complex kinds of processes can be modelled in this way based on fundamental equations and laws of physics. In the cosmological domain, we use Einstein's equations of general relativity to describe the evolution of the early universe, the creation of gravitational lenses, and the effect that dark matter has on the behaviour of normal matter. We cannot obviously experience these ancient times or vast scales ourselves, so we use the relevant

equations of physics and motion to create models using supercomputers. These models, like the Millennium Universe Simulation in 2005, can be played back as movies so that human observers can 'see' what was going on, and whether the mathematical models look anything like the real world astronomers observe as they look outwards into space.

Cosmology and astrophysics are relatively straightforward because they deal with matter operating through gravity to form galaxies, stars and planets among other structures. But what happens when we can no longer visualize the objects under study? This is the challenge we have in the quantum world where we study things we call particles, but these do not resemble similar objects in the macro world. This is a huge stumbling block. Because of the way these new objects

An example of the 3D structure of the universe at a given moment in cosmic history simulated by the Millennium Universe Simulation. The threads are the network of galaxies surrounding vast empty voids. The frame is 253 megaparsecs (824 million light years) in distance across. The universe is a mixture of dark matter and ordinary matter. Dark matter collapsed under its own weight to form vast halos (bright yellow) which sucked in normal matter to form visible matter, such as galaxies.

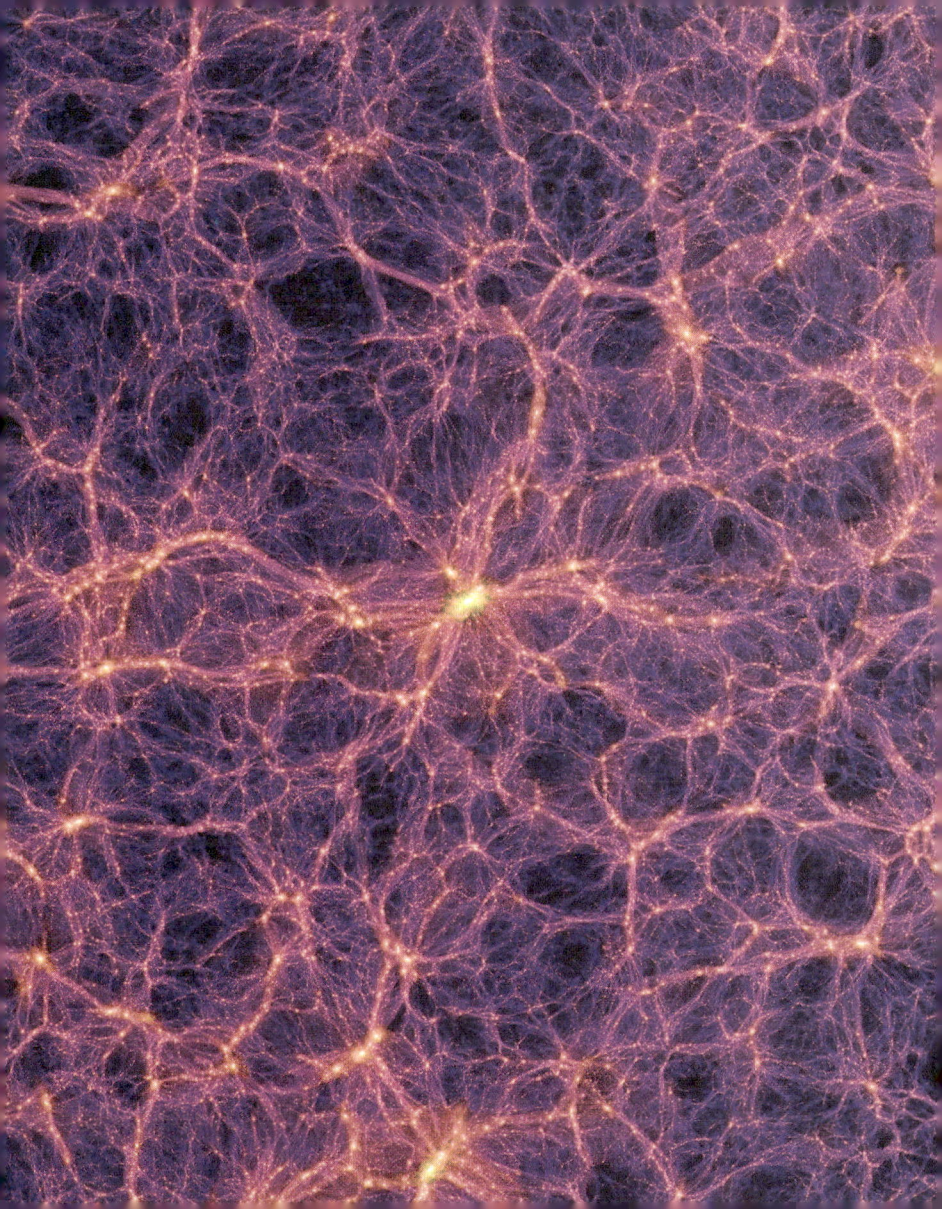

behave, not a few physicists have remarked that quantum mechanics is nearly impossible to understand using the tools of common sense, internal visualization, or any other approach that has served us so well for millions of years.

As the astrophysicist, John Gribben writes: *'In the world of the very small, where particle and wave aspects of reality are equally significant, things do not behave in any way that we can understand from our experience of the everyday world... all pictures are false, and there is no physical analogy we can make to understand what goes on inside atoms. Atoms behave like atoms, nothing else.'*

This is not a book that will teach you all the essentials of quantum mechanics. Instead, in this chapter I will introduce ideas that only relate to time and its observation at the quantum scale. Whereas macroscale physics at the human scale and beyond describes the collective behaviour of billions upon billions of atoms and the collective action of trillions of electrons in the properties of electricity, quantum physics involves the study of individual elementary particles, how we observe them, and the rules they follow in space and time. There is, conceptually, no reason to expect that the rules of one domain should apply when we examine another, far smaller, domain. For example, we use temperature to gauge the energy in our environment, but temperature is a collective property of matter that utterly vanishes and becomes meaningless at the level of individual particles and atoms.

Even more telling is Niels Bohr's comment that atoms and electrons are not like tables and chairs – instead they are an entirely different order of reality that plays by different rules. Werner Heisenberg, one of the founders of modern quantum mechanics, took this idea even further and thought that electrons only actually exist at the point of interaction, like a person walking in a dark park illuminated here and there by lamp posts (interactions); the path taken to get to each lamp post is invisible and unobservable to us. According to physicist Carlo Rovelli in his book *Reality Is Not What It Seems*, an electron does not travel *through* space along a path at all, but appears at the instant of an interaction almost literally from out of nowhere. This means that reality is reduced to interactions. It is only in interaction that nature draws a world. Because interactions are related to events along worldlines, it is the worldlines that define reality and nothing else.

It is commonplace knowledge that atoms are fashioned from electrons, protons and neutrons and from this we get

the Periodic Table of the Elements. We also understand that light can be described as an electromagnetic wave, but that in some kinds of experiments it behaves as a discrete particle we call a photon. This leads to the concept of particle-wave duality in quantum physics. You can fashion theories using one description or the other but not both simultaneously. Even as a particle, an elementary electron has no surface or volume. It is not a miniscule sphere of matter with electric charge on its surface. Nevertheless, physicists do internalize quantum particles as tiny spheres of energy like effervescent marbles. Our brains understand marbles. They do not understand quanta of energy.

Another thing we have to avoid imagining is that within the atoms, electrons orbit the nucleus like miniature planets. Instead, their wave-like character requires that

Werner Heisenberg proposed that electrons only exist as real particles at the point where they interact with something else, or are being observed by an instrument. This is like a pedestrian existing only under each lamp post. We cannot know where the pedestrian was as they walked in the darkness between the lamp posts. In a real sense, the person ceases to exist except under each lamp post.

they behave like standing waves of energy; the kinds of waves you create in a rope tied at one end and shaken at the other. Sometimes when you try to observe this electron it will be in one location within the atom. At another time you will detect it at another location. When you combine all these position measurements together, you see the electron spread out within the atom according to a probability distribution. The details of this probability distribution can be calculated exactly by using the fundamental equations of quantum mechanics such as Schrödinger's equation. This calculation is an elementary exercise for undergraduate physics majors, who become proficient in its use after a few months of training.

Does this probability distribution really exist? Can you take a picture of it? In fact, in 2013, physicist Aneta Stodolna, of the FOM Institute for Atomic and Molecular Physics in the Netherlands, along with Marc Vrakking at the Max-Born-Institute in Berlin, Germany, and other colleagues in Europe and the US, did just this. They made 50,000 measurements of the position of the electron in a carefully prepared hydrogen atom by using the electron's interaction with photons and then calculating backwards where the electron had to be in the atom at the moment of the collision. Combining all these measurements they 'took a picture' of the electron's probability cloud.

These and many other experiments confirm that we have a deep understanding of quantum matter at the level of being able to make detailed predictions of what to expect from nearly any interaction between quantum particles and electromagnetic radiation. However, this mathematical insight hides a number of fundamental features of our physical world that are still being studied. Each of them leads to either a paradox or a deep and puzzling mystery.

The first feature is called 'locality', which is a blend of the principles of quantum mechanics and Einstein's relativity. It means, basically, that every object can be influenced only by events that travel no faster than the speed of light. Interactions between particles are not instantaneous but have delays that are set by the speed of light. For example, Newton's theory of gravity is a non-local theory because the influence of gravity was assumed to be instantaneous. In quantum mechanics, some of the earlier ideas for how to eliminate probability from the calculations were to imagine what were called 'hidden variables' or influences that made the calculations exact. We only use probability because we can't directly observe these hidden interactions. In 1964,

physicist John Stewart Bell came up with a simple experiment to test whether quantum mechanics is local or non-local. The result of many different tests of quantum mechanics has been that quantum mechanics is indeed a local theory consistent with relativity and having no hidden variable improperness within it to remove its probabilities. So, the indeterminacy of quantum mechanics cannot be fixed by adding hidden variables to make it deterministic. Indeterminacy is a feature of quantum mechanics and not a flaw.

The second important, and curious, feature of quantum mechanics is called entanglement. The recognition of this effect is a result of a famous thought experiment proposed by Albert Einstein, Boris Podolsky and Nathan Rosen (EPR) in 1935. What it says is that, when you prepare a quantum system in a single state, you can separate the components of that state to huge distances and they still remain in a single quantum state. But if you measure very rapidly the properties of one component of that state, the other component located

An electron is represented by a cloud of possible locations in space determined by its quantum mechanical wave function, much like the concept of the 'star cluster M13' is represented by a cloud of stars in space.

far away will suddenly and instantaneously have the predicted state. As a concrete example, some particle decays produce two photons travelling in opposite directions: call them A and B. The polarizations of the photons can be either left or right (up or down) so the two possibilities are A = Up, B = Down or A = Down and B = Up. The sum of the polarizations have to be exactly zero (A + B = 0). In the EPR experiment, if you measure the polarization of photon A to be Up, you will instantly know that the polarization of photon B is Down or vice versa. This experiment was finally carried out in the early-1980s by Alain Aspect over distances of a few metres. The data was used to test Bell's inequality and Aspect and his team found that the photon spins were indeed correlated as predicted, even though they were far apart.

In 1997, physicist Nicolas Gisin and his team in Geneva performed the EPR experiment across 10 km (6.2 miles) using pairs of photons in entangled states sent separately down two fibre optic cables. The result was that measurements on each end showed that the spins of the photons were random, but that the combined system always had one end having the opposite spins at the level of individual photons. It was not possible for the spin information in each pair to travel at the speed of light to align them, so quantum mechanics is indeed a non-local theory where information and influences seemingly travel faster than light.

Not only does the EPR experiment work for single particles but in 2023 it was tested on a collection of 700 rubidium atoms in what is called a Bode–Einstein condensate. The rubidium atoms were prepared in one entangled state, then carefully split apart into two group separated by 100 microns. When the spins of one group were measured, the spins of the second group were complementary to those of the first group in a way that could not be accounted for by random chance. Because this information travels faster than light, entangled systems are quantum systems that are non-local.

The last of these features, called realism, is that elementary particles have properties that are fixed and determined *before* the particle is observed. As Einstein famously notes: *'I like to think that the moon is there even if I'm not looking at it.'* This aligns with our common-sense notion that things like planets, baseballs and even bacteria have well-defined properties like mass, size and even colour that are permanent features even when we are not looking at them. Apparently, this is not the way the quantum world works. This insight about the quantum world, that it violates realism, was not

something we stumbled upon recently but was baked-in to quantum mechanics at the outset. Heisenberg, one of the developers of quantum mechanics in the 1920s, said of this situation that *'[T]he atoms or elementary particles themselves are not real; they form a world of potentialities or possibilities rather than one of things or facts.'* Tests of realism were conducted simultaneously with tests of locality using Bell's inequalities and the result is that the EPR experiment does not support the idea that photons 'know' what spins they are supposed to have before they are measured. The set-up of the measurement process itself determines what properties they take on.

The conclusion from these experiments, and still an active area of research today, is that quantum mechanics is a non-local theory and that it also violates realism. How this all impacts our understanding of time is significant. Entanglement creates quantum states that are indivisibly connected to each other but in which the physical properties of the individual parts – be they electrons or photons – are not fixed. They seem to depend on the outcome of an observation, measurement or interaction.

There is an interesting parlour game suggested by John Wheeler that is a variant on the 21 Questions game. Instead of having an object firmly in mind at the outset, the answers to the questions themselves narrow the possible objects to the one in question. It isn't until the last answer is provided that the object, unknown to any of the participants, finally comes into existence. If quantum mechanics and Wheeler's 'observer participatory universe' are correct, it is the act of observation, or measurement, or simple mindless interaction, that brings features of objects in our universe into existence. Could time be like that?

THE SEARCH FOR TIME IN A BOTTLE

Modern ideas about time suggest that it only exists to observers inside our universe. Outside 'observers' exist in a timeless state. In real terms, time is trapped inside the 'bottle' of our universe.

In the macroworld, we may have an equation that specifies the position of a tennis ball in space for all possible trajectories the ball could traverse, given the energy and momentum of the hitter's impulse and the strength of the force of Earth's gravity. It is an infinitude of trajectories. To find a specific trajectory for a particular tennis game I am watching, I have to specify for the observed tennis ball the exact energy and momentum and Earth's acceleration. Then I get an equation for the parabolic path through three-dimensional space. In quantum mechanics, you work with a quantity called the wave function. It is called a function because, in fact, it is itself an equation that specifies how a particle's probability (actually the square root of the probability) will be determined for all possible positions in 3D space, and for all times. But you don't get just one equation – you get an infinite number of equations. Each one represents a possible quantum state for the particle defined by the values of a set of observable 'quantum numbers'. To determine which equation (quantum state) is being observed, you have to experimentally determine the quantum numbers to get the specific equation that applies to your observation at the moment.

Now in both the case of the tennis ball and the quantum particle, all we have is an equation for the position of the particle or ball as a function of time. But humans do not see the entire path of the tennis ball all at once. We see it at a specific time in its trajectory because our brains grab 100-millisecond snapshots of the world and build models from this about where things are and how they are changing in space. Similarly for a quantum particle, time only comes into the picture when we make an observation of where the particle is at a given moment. Prior to this, both the tennis ball and the quantum particle were in a timeless state analogous to the worldlines drawn for the histories of particles in relativity. So now we see that in the quantum world that the instant of the interaction of a particle with another particle or system, which could involve a measurement in some laboratory, or simply the collision of a particle with a photon, is unique in the worldline of the particle. It occurs at a specific proper time.

If no observation is ever made of this particular electron in this particular hydrogen atom, its location in space is a timeless cloud of possible positions. In essence, the electron becomes an eternal object with a cloud that manifests all possible positions of the electron, not just for a particular moment but for all time. This analysis doesn't help us very much if we are trying to understand the origin and nature of time. To go from the description of the specific quantum state of a system described by the timeless wave function to a specific prediction of where the electron is in space, you have to make an observation to specify the instant of the observation, say t = t(now). We have simply moved the behaviour of the particle in time to the passing of time in the hands of the observer, which decided when t = t(now) was to be. Is there any other aspect of the quantum world where the nature of time is laid bare? There appears to be an interesting situation that involves entangled quantum states.

We saw how the entanglement of two particles was used to explore the EPR experiment and deduce from it that quantum mechanics displays what is called non-local realism. Information between the two entangled systems seems to violate special relativity and the speed of light being the maximum speed by which information and influences can be transmitted through space (non-locality). It also demonstrated that the two entangled states do not know what their properties are until after they are observed or measured (realism). So, the quantum world is both non-Local and non-Real. As it turns out, there is something weird with time as well. Imagine that the two entangled states carried their own clocks and that the external observer also had a clock. Do all of these clocks mark the same times? Apparently not.

In the 1960s, it was a popular theoretical activity to try to combine quantum mechanics with general relativity: the two great theories of the quantum and cosmological worlds. At the forefront of this research were physicists John Wheeler and Bryce DeWitt. They conjectured that to combine these two systems, you had to use a common language in which the entire universe was displayed as a series of superposed quantum states defining the complete geometry of spacetime. What we observe is this superposition, which on the cosmological scale averages out to the simple spacetime geometry we observe, but at the quantum scale of 10^{-33} cm is constantly fluctuating between many neighbouring states creating a foam-like spacetime. The second part of their idea was in direct parallel with the foundations of quantum mechanics. As we saw on page 96, there was a mathematical space, called superspace, in which each point in the space was a complete three-dimensional space geometry for the entire universe from the quantum level to the cosmological scale. Our particular universe is represented by a path that connects a small number of these states, and for various reasons, represents the path of maximum probability. They produced an equation called the Wheeler–DeWitt equation that computes exactly what that path should look like, but the catch was that this equation made absolutely no reference to any time variables. Apparently, superspace and its paths were beyond time itself. By some process, the physical attribute we call time does not come from a simple model of 'quantum cosmology'. This discovery became known as the 'problem of time'. Was there any way to test this idea? How could one possibly step outside the universe and our own spacetime and set up the necessary external clocks for comparison? As it turned out there was a way to do this, at least theoretically.

Physicist Don Page at the University of Alberta and William Wootters at Williams College looked at this problem in 1983 using standard techniques in quantum theory. They, mathematically, split the universe in half and imagined each half as part of an entangled quantum state. Taking advantage of the entangled state principle that one of the states would contain the opposite information of the other, each of the states were free to evolve according to their own internal clock times, but as viewed from an external space, the times would cancel out exactly. The external observer in superspace would observe no time passing within the combined entangled state. So, time and entanglement were related phenomena and this is how the Wheeler–DeWitt problem of time could be resolved. From outside the path of the universe through superspace, there was no time, but from inside the path and between the individual quantum states for the universe, entanglement allowed something like a time-like phenomenon to exist.

This idea that no external time exists but is something created by internal configurations of matter, was finally put to the test in 2013 by physicist Ekaterina Moreva and her colleagues at the Istituto Nazionale di Ricerca Metrologica in Italy. They created pairs of entangled photons using an argon laser passed through a series of beam splitters and optical plates that rotated their planes of polarization. They adjusted the optics through specially designed quartz plates so that the pairs of photons had opposite polarizations. In the first experiment, they created an internal clock effect by measuring the polarization of one of the photons against the other as the plate thickness changed. By measuring the polarization of one of the photons in the pair, the observer becomes entangled with it. This polarization is then compared with the polarization of the second photon. The difference is a measure of time, and the internal clock between the two entangled states behaved just as Page and Wootters had proposed it would. Next, the experiment was run again to mimic the experience of an outside 'super-observer', but this time the observer only measures the global properties of both photons by comparing them against an independent clock. The super-observer could never look at the internal clock because to do so they would become entangled with it and record the passage of the internal clock time. Moreva looked at the two entangled photons and this time measured their combined polarization. The result after many repeated measurements was that the combined polarization was always zero, meaning that the polarization 'clock' had stopped for the super-observer watching the entangled system. This is the perspective of the Wheeler–DeWitt universe. This experiment confirms that time is an emergent property of an entangled system as perceived by 'clocks' within a universe, but not by an external clock as for the Wheeler–DeWitt system. What this also means is that, very literally, our entire universe is a single quantum object within which things are entangled and from this entanglement, time itself emerges.

Experiments such as these at the quantum level suggest that the block universe perspective of relativity, which we discussed in Chapter 7, cannot be correct. Although the past can be reconstructed from deterministic calculations in general relativity and fixed records of stored information (photographs etc.), the future is closely determined by probabilities and principles of indeterminism found in quantum mechanics. The 'present' is when the quantum mechanic probabilities of the future become 'crystallized' into the certainty of the past in the form of an event that is created from some interaction. Physicist George Ellis in 2009 proposed this as the concept of the crystallizing block universe. So, time may not exist in the Wheeler–DeWitt superspace external to our universe, but how does entanglement actually create time itself? The answer seems to be that, the more entanglement a system has, the more entropy it has, and this entropy increase is in the direction of the passing of time.

An intriguing set of papers by physicist Seth Lloyd at Harvard University in 1984 showed that entanglement is actually how systems evolve into an equilibrium state. Over time, the quantum states of the member particles become correlated (entangled) and shared by the larger ensemble.

An artistic rendering of entangled particles. Entanglement may create time between them, which can be internally measured; but to an outside observer not entangled with the particles, there is no time involved.

This direction of increasing correlation goes only one way, dictated by the increase of entropy, and establishes the 'arrow of time' on the quantum scale. Cosmologically, it is apparent from a variety of measures that events are connected in time in a definite past-present-future order. This obtains from a variety of biological, thermodynamic and cosmological measures of closed systems changing their states generally from order to disorder following the second law of thermodynamics. These are called the arrows of time, a term popularized by Sir Arthur Eddington in his 1928 book *The Nature of the Physical World*. We always see systems that evolve from an ordered state to a disordered one such as an ice cube melting to water, but we never see water suddenly forming a more organized ice cube. This change in entropy in entangled systems is now seen by physicists to be synonymous with, and a proxy for, the passing of time.

LIVING WITH A SHORT SPACE OF TIME

A possible artistic rendering of the nature of quantum spacetime. According to loop quantum gravity, space only exists at discrete nodes in a network in which the links in the network are non-physical bits of information that follow causality. Between the nodes and the links there is pure nothingness.

Quantum mechanics has given us deep insight into the aspects of three of the fundamental forces of nature; electromagnetism and the strong and weak nuclear forces. These are now described by powerful theories that combine special relativity and quantum mechanics into descriptions of the forces called quantum field theories. Within this rubric, we can perform impressively detailed calculations borne out by direct observation and measurement in some cases to nine-decimal-place accuracy. Collectively the field theories for these three forces are called the Standard Model and describe how a fundamental set of elementary particles interact via these forces. In fact, two of these three, electromagnetism and the weak nuclear force, have already been 'unified' into what is called electro-weak theory, which 50 years ago predicted the existence of the Higgs boson, discovered at the Large Hadron Collider in 2012. However, the Standard Model has between 18 and 26 fundamental constants that cannot be derived from within the Standard Model. The hope and expectation is that a future Theory of Everything (TOE) will greatly reduce these dozens of parameters to perhaps only a handful; and that in relativity, the speed of light and the constant of gravity, and in quantum mechanics Planck's constant, will play a central role among these truly fundamental constants. The interesting question is whether the three dimensions of space and the one dimension of time are also free parameters that have to be experimentally determined like the parameters in the Standard Model. What, for example, do TOEs look like if spacetimes have two dimensions of space and three dimensions of time?

Many physicists today (in fact since *c.* 1930) are attempting to find a way to build such a TOE, but the challenge is enormous. We already saw one such attempt in the 1960s by Wheeler and DeWitt to describe general relativity in quantum terms. Their idea was that at any instant in time, our three-dimensional space was in one of an infinite number of possible quantum states. They even proposed an equation similar to Schrödinger's equation to describe how one cosmological state changes into another. However, unlike Schrödinger's equation the Wheeler–DeWitt equation did not include time at all. In fact, their approach says absolutely nothing about the kinds of particles and matter responsible for creating the various three-dimensional quantum states of the universe. In the 1980s and 1990s, string theory was proposed as a way of describing elementary particles and fields in terms of geometric objects called strings. A number of promising discoveries were made that allowed gravity to be described as a quantum field composed of closed strings, while Standard Model particles and fields were open strings. But string theory is a theory that presupposes spacetime already exists, which would not be the case if it were a theory that truly unified gravity with the other forces. For this reason, string theory is called a background-dependent theory. What the TOE has to be is a background-independent theory in which spacetime is created from within the theory itself. One of the most recent ideas for a candidate TOE, now over 30 years old, is called loop quantum gravity (LQG). Although it, too, seems to have no place in it for actual elementary particles, its practitioners are hopeful that it may at last provide an explanation for why space and time exist at all.

Loop quantum gravity, developed in the 1990s by Lee Smolin and Carlo Rovelli among others, proposes that spacetime can be resolved into individual nodes that have Planck-scale volumes of $(10^{-33} \text{ cm})^3$ or 10^{99} cm^3. These nodes define where three-dimensional space exists, and are connected to each other by links like a tinker-toy. These links, however, do not exist in space, but nevertheless carry information related to the quantum area $(10^{-33} \text{ cm})^2$ to be assigned to each node. Three-dimensional space is represented by networks of these nodes and links called spin networks. But spin networks can change, and these changes form a four-dimensional network called a spin foam. This spin foam is a 3+1-dimensional structure where the '1' represents the direction along which the changes between spin networks (quantum states of 3D space) occur.

The thing to remember about LQG is that the nodes, which represent quanta of space the way that photons represent quanta of the electromagnetic field, do not exist in space. They *are* space. They have no place to be because they *are* that place. They only contain one snippet of information and that is which other quanta are near to them. This is how they are defined; not by where they are in relation to a pre-existing space, but where they are in relation to their neighbours. Then we have the links between them.

The links between the quanta of space do not exist in space or in time. They are nowhere at all and are a purely mathematical tool to define the network of nodes. Physical space is, as Rovelli describes it, '...*the fabric resulting from the ceaseless swarming of this web of relations. The lines themselves are nowhere. They are not in a place but rather create places through their interrelations. Space is created by the interaction of individual qualia of gravity.'* Space itself no longer exists as the starting point for the theory. Space comes into existence out of the network interrelationships. Time only serves as a means of counting the number of changes that have taken place between one configuration, or quantum state of space, and the next. Taken in their totality, the changes between one quantum state and the next occur by means of a sequence of specific 'moves' that spawn new nodes and relationships. This ensemble is called a spin foam and exists in a four-dimensional arena in which the three-dimensional spatial quantum states are stacked along a fourth dimension or axis that records the specific changes between the quantum states, much like the pages in this book are stacked between the book's covers. You are entirely free to read the pages in this book in any order you want, but there is an underlying timeless order suggested by the linguistic logic that is based on a cause-and-effect relationship among these pages. If you instead sliced the book perpendicular to the pages you would only find a jumble of words in each slice that did not combine to yield a coherent story.

What is remarkable about spin foams is that, although they seem to be purely spatial in four dimensions, in fact if time is defined as 'change', time emerges from changes in spin networks along the fourth dimension. In fact, like Barbour's concepts of time capsules, these three-dimensional networks contain within them information about their past states. The reason for this assignment of the axis of 'change' to 'time' has to do with one of the most important features of relativity called causality. If you select any of the other three dimensions and ask how the spin foam is organized along them, like slicing a book perpendicular to its pages, you will only recover a cacophony of randomly related events and states, but along the fourth dimension, the organizing principles of these changes is laid bare. That's why this fourth dimension, which encodes changes in a logical manner based on causality, is so special.

With causality, we can connect events together in a specific order, but this order does not have to be specifically a time order. Think about a drawing of your family tree. The lines you draw between your grandparents and parents and you are intended to be cause-and-effect indicators, not physical lines in space. They do form a very crude timeline. In LQG, causality lets us assign changes in a spin foam to sequences that we interpret as a time ordering. According to Lee Smolin, '*There is no such thing as a clock existing outside*

the network of relationships. Time is described only in terms of changes in the network of relationships that describe space.' Any clock you try to create would be embedded within the spin network and spin foam as its own small collection of quantum states. This leads to the idea that what we call time is actually something else entirely and actually emerges from spin foams at the Planck scale where these spin foams exist.

We saw in the previous chapter on entanglement that this idea that time is an emergent phenomenon from within our spacetime network, was proposed in 1983 by Don Page and William Wootters as a dramatic solution to the fundamental origin of time based on a property called quantum entanglement. *'A static, entangled state of two photons can be seen as evolving by an observer that uses one of the two photons as a clock to gauge the time-evolution of the other photon. However, an external observer can show that the global entangled state [of the two photons] does not evolve.'* It was confirmed by a set of experiments conducted in 2013 by physicist Ekaterina Moreva at the Istituto Nazionale di Ricerca Metrologica.

In 1983, James Hartle at UC Santa Barbara, and Stephen Hawking at the University of Cambridge proposed a different origin for time in the Big Bang in their 'no boundary condition'.

The Big Bang was a quantum mechanical 'tunnelling' of one of the four space-like dimensions into a time-like dimension, which according to Alexander Vilenkin at Tufts University caused eternal inflation. Although the initial state (what Vilenkin called 'nothing') was a pure-space condition in many dimensions, once one of the dimensions emerged as the direction of a causal, past-future succession of spatial states, through the deus ex machina mechanism of quantum tunnelling. The Big Bang occurred and progressed in the direction of increasing entropy defining the arrow of time.

An even more ambitious idea called geometrogenesis has been proposed by physicist Daniele Oriti at the Max Planck Institute for Gravitational Physics. Spacetime may just be a phase transition from more primitive constructs that are not even space-like or time-like. We arrive at the current 3D space and 1D time through a process similar to condensation of gas into a liquid. The transition would have been in the direction of increasing entropy. Although not directly related to the evolution of time, the current accelerated expansion of the universe under the action of dark energy has led some physicists to imagine that dark energy may be signs of our universe 'growing' a fourth spatial dimension as the process of geometrogenesis continues even today.

THE BOTTOM LINE

An artistic rendering that might convey some of the mystery of the birth of spacetime. It may have been an event in which four dimensions of space transformed into three dimensions of space and one dimension of time.

Our experience of time may be subjective and limited to an immediate sense of Now, but on the cosmic scale, time seems to be a feature of entangled relationships between objects in space and not a feature of anything 'outside' our universe. Its directionality is a consequence of the increasing entropy of an expanding universe since the Big Bang, an initial condition not derivable from known physics. Although time's arrow precludes 'remembering the future' the low entropy of past configurations of matter allow for fossil records to exist from which past events can be recovered and stitched into a consistent story. Meanwhile, the causal links between events along the numerous worldlines that define our physical four-dimensional spacetime takes us all the way back to the Big Bang itself using many different kinds of clocks. During the Planck era, it may be that time emerged from a pure space-like condition through a quantum tunnelling event out of 'nothing' that jump-started the Big Bang. Once time became separate from the other three space-like dimensions, our universe emerged from the Planck era at 10^{-43} seconds. The past-future directionality imposed on this process as an initial condition allowed inflation in our particular bubble universe to occur, and our current, cosmic Now appeared after 13.8 billion more years.

But why are we living in a cosmic Now occurring 13.8 billion years after the Big Bang, or a personal Now for you occurring in the year 2025 CE and not 3658 CE or 1165 BCE? Along my worldline and in my Now of 2024, this remains an outstanding mystery of time and the circumstances of our very existence in this universe. Along your worldline, in your Now of 2025, there is a chance that physicists may already have discovered the answer!

Why did I start this book with so much talk about brain research? Well, it is the brain, after all, that tries to create ideas about what you are seeing based on what the senses are telling it. The crazy thing is that what the brain does with sensory information is pretty bizarre when you follow the stimuli all the way to consciousness. In fact, when you look at all the synaptic connections in the brain, only a small number have anything to do with sensory inputs. It's as though you could pluck the brain out of the body and it would hardly realize it needed sensory information to keep it happy. It spends most of its time 'talking' to itself.

The whole idea of space really seems to be a means of

representing the world to the brain to help it sort out the rules it needs to survive and reproduce. The most important rule is that of cause and effect or 'if A happens then B will follow'. This also forms the hardcore basis of logic and mathematical reasoning!

But scientifically, we know that space and time are not just some illusion created by our brain merely to organize sensory information, because objectively they seem to be the very hard currency through which the universe represents sensory stimuli to us. How we place ourselves in space and time is an interesting issue in itself. We can use our logic and observations to work out the many rules that the universe runs by that involve the free parameters of time and space. But when we take a deep dive into how our brains work and interface with the world outside our synapses, we come across something amazing.

The brain needs to keep track of what is inside the body, called the 'self', and what is outside the body. If it can't do this infallibly, it cannot keep track of what factors are controlling its survival, and what factors are solely related to its internal world of thoughts, feelings and imaginary scenarios. This cannot be just a feature of human brains, but has to be something that many other creatures also have at some rudimentary level so that they too can function in the external world with its many hazards. In our case, this brain feature is present as an actual physical area in the cerebral cortex. When it is active and stimulated, we have a clear and distinct perception of our body and its relation to space. We can use this to control our muscles, orient ourselves properly in space, walk and perform many other skills that require a keen perception of this outside world. Amazingly, when you remove the activity in this area through drugs or meditation, you can no longer locate yourself in space and this leads to the feeling that your body is 'one' with the world, your self has vanished, and in other cases you experience the complete dislocation of the self from the body, which you experience as 'out of body' travel.

What does this have to do with space in the real world? Well, over millions of years of evolution, we have made up many rules about space and how to operate within it, but then Einstein gave us relativity, and this showed that space and time are much more plastic than any of the rules we internalized over the millennia. But it is the rules and concepts of relativity that make up our external world, not the approximate common-sense ideas we all carry around with us. Our internal rules about space and time were never

designed to give us an accurate internal portrayal of moving near the speed of light, or functioning in regions of the outside world close to large masses that distort space.

But now that we have a scientific way of coming up with even more rules about space and time, we discover that our own logical reasoning wants to paint an even larger picture of what is going on and is happy to do so without bothering too much with actual (sensory) data. We have developed for other reasons a sense of artistry, beauty and aesthetics that, when applied to mathematics and physics, has taken us into the realm of unifying our rules about the outside world so that there are fewer and fewer of them. This passion for simplification and unification has led to many discoveries about the outside world that, miraculously, can be verified to be actual objective facts of this world.

Along this road to simplifying physics, even the foundations of space and time become players in the scenery rather than aloof partners on a stage. This is what we are struggling with today in physics. If you make space and time players in the play, the stage itself vanishes and has to somehow be recreated through the actions of the actors themselves. *That* is what quantum gravity hopes to do, whether you call the mathematics loop quantum gravity or string theory. This also leads to one of the most challenging concepts in all of physics...and philosophy.

What are we to make of the ingredients that come together to create our sense of space and time in the first place? Are these ingredients, themselves, beyond space and time, just as the parts of a chainmail vest are vastly different than the vest that they create through their linkages? And what is the arena in which these parts connect together to create space and in particular time itself?

The idea of emergence, which is the current way to look at time, is that elements of nature come together in ways that create new objects that have no resemblance to the ingredients, such as evolution emerging from chemistry, temperature emerging from atomic physics, or mind emerging from elementary synaptic discharges. Another striking example of emergence is in flocks of starlings called murmurations. Each starling flies according to a simple rule – 'stay as close to your neighbour as possible' – but from this simple rule, flocks create dynamic and complex clouds with constantly changing form and geometry.

If the current research directions are on the right track, apparently, time and space may also emerge from ingredients still more primitive, which may have nothing to do with either

time or space. The true nature of time, itself, may simply be one of the many fundamental 'constants' of nature that came together with just the right values and properties to create our physical universe out of the cauldron of the Big Bang and an infinitude of random events and states. We may never know from within our universe why time or even the universe exists at all, but we are of course grateful that they do. In the words of the ancient Roman philosopher Seneca in his book *Natural Questions*:

'The time will come when diligent research over long periods will bring to light things which now lie hidden…. There will come a time when our descendants will be amazed that we did not know things that are so plain to them…. Many discoveries are reserved for ages still to come, when memory of us will have been [worn away].'

The emergence of time from space may share something in common with the emergence of structure from the murmurations of starlings as they fly, but follow the simple rule of keeping close to their neighbours.

INDEX

PICTURE CREDITS